Collins *gem*

Scottish
Birds

Valerie Thom

Valerie Thom worked as an agricultural adviser and in conservation education before becoming a freelance writer on the Scottish countryside. She wrote the Scottish section of the *Macmillan Guide to Britain's Nature Reserves* (1984), *Fair Isle: an Island Saga* (1989) and *Birds in Scotland*, published in 1986 to mark the Golden Jubilee of the Scottish Ornithologists' Club, of which she was an Honorary President.

HarperCollinsPublishers
77–85 Fulham Palace Road
Hammersmith
London W6 8JB

First published 1994 as *Collins Scottish Birds*
This Gem published 2006

10 9 8 7 6 5 4

ISBN-10 0 00 720769 7
ISBN-13 978 0 00 720769 5

Illustrations by Norman Arlott
Gaelic names kindly provided by Calum Campbell

Thug Comhairle nan Leabhraichean tabhartas barantais airson gun cuirte ainmean Gàidhlig nan eun dhan leabhar

Printed and Bound in China by Leo Paper Products

CONTENTS

ABOUT THIS BOOK

This book aims to help you recognise most of the birds you can expect to see in Scotland. It does not cover birds which visit only occasionally, or which occur in small numbers or are difficult to identify. Instead it concentrates mainly on the commoner species, plus a few rarer ones included because they are especially 'Scottish' or are of particular interest. Each species description highlights the features which help in identifying a bird and gives information on the habitat(s) occupied by each species, the season during which it is present, and the areas in which it occurs.

A summary of the most common habitats is given on pages 9–15. Birds are very mobile creatures and often make use of different habitats at different times of year; it is seldom possible to say that a particular species occurs only in a specific habitat. Some Scottish breeding birds move to a different habitat in winter, or move south, which means that not all individuals of a species represented throughout the year are actually permanent residents. Some species are more likely to be seen when on migration in spring and autumn than at any other time. Knowledge of preferred habitats, along with identifying features, flight patterns and song, will help you to recognise birds when in the field.

SOME GENERAL HINTS ON BIRD IDENTIFICATION

Many different aspects of a bird's appearance and lifestyle can assist in its identification. Some are obvious:

(1) Its size – larger, smaller or similar to a familiar bird.
(2) Its proportions – relative length of neck, legs and tail, stocky or slender body.
(3) The shape of its wings and tail – wings pointed or rounded, tail long or short, forked or rounded.
(4) The size and shape of its bill – small and slim for picking out insects, short and strong for seed cracking, long and slender for probing in mud, or hooked for tearing flesh.
(5) Its general colour (may vary with sex, age or season) and the position and colour of any distinctive markings – especially patches on wings, rump, tail, stripes on head or bars on wings.

Other characteristics are less obvious:

(6) The way it moves – does it walk, hop, run, crouch, climb; flies – fast and direct, in loops, hovering or soaring, returning to the same perch; dives into water – plunging head or feet first, sliding under the

surface or giving a quick jump first; uses its tail –
cocking, flicking, fanning or wagging.

(7) The way it feeds – picking off plants or the ground,
catching flies or other birds in flight, pouncing on
prey, probing deeply or shallowly.

(8) The season, and the habitat the bird is using – many
species are in Scotland for only part of the year and
some occur only in very specific types of habitat.

And, of course:

(9) The noises it makes – some birds are elusive and
difficult to see but can be readily identified by their
song or calls – with practice! To recognise calls and
songs, go out with an experienced birdwatcher and
listen to good recordings. Start in early spring, so
that you are familiar with the resident species' songs
before the summer visitors arrive.

A CODE OF CONDUCT FOR BIRDWATCHERS

Help the birds by keeping disturbance to a minimum

- avoid scaring birds at the nest – eggs or young may become chilled, or be taken by predators if the nest is abandoned
- take extra care where ground-nesting birds breed – well camouflaged eggs can be trodden on
- do not disturb feeding birds unnecessarily in very cold weather – they may be struggling to survive
- rare breeding birds are at risk from egg-collectors – only visit protected haunts

Help to conserve birds by protecting their habitats

- do not start fires on heaths or in woodlands
- keep to paths wherever possible
- do not drop litter – some forms of litter can be lethal to birds

Respect the law and other people's rights

- abide by the bird protection laws at all times
- always obey the Country Code
- respect the rights of landowners and always get permission if you will be going onto private property

(Based on the code produced by the RSPB)

HABITATS

Gardens, Parks and Buildings

Many common garden bird species have become accustomed to the benefits of living close to man – the regular supply of suitable food and water, the shelter and natural food obtainable on, under or around cultivated trees and shrubs, and the varied nesting and roosting sites provided by buildings. Most people start their birdwatching in a garden or park, where many of the birds are tame enough to give good close-up views for studying identification characteristics.

The number and variety of birds that will visit is determined more by the nature of the plant cover and the food supply than by the size of the garden or park. In general, gardens with a mixture of trees, shrubs and open ground will support the greatest variety of birds, and a pond and berry-bearing shrubs are definite attractions. Many garden birds can be attracted with seeds, fat or scraps at bird tables or in nut baskets.

Several bird species, such as house martins and swifts, select buildings as their preferred nest site. Less usual building-nesters include kittiwakes, kestrels and even a few peregrines on the ledges of buildings, and herring gulls and oystercatchers on flat roofs. Buildings also provide roosting places, sometimes attracting large numbers of birds in winter, when it is important that body heat is maintained.

Farmland

Farmland provides a variety of food resources which suit a wide range of species. On intensively cultivated arable land, both the crops and the season influence the birds that will be seen. In autumn both spilt grain and weed seeds on stubble are frequented by a variety of species from pheasants to finches. Potato fields are used as resting areas, and ploughing brings many gulls inland to search for worms. If arable fields have hedges round them and patches of woodland nearby, there is likely to be a good variety of breeding birds, including woodland and hedgerow species.

Rough grassland is common on farms in Scotland's glens, and is important for breeding waders and other ground-nesting species. Some of the species that feed over grassland do so by probing into the soil in search of insects, worms and other small creatures. For nesting birds, rough or damp grassland with tussocky growth for cover is better than heavily grazed pasture.

A characteristic of most crofting areas is the mosaic of grassland, cropped ground and fallow land, often closely associated with stretches of machair, bog and short-heather moorland. This mixture of vegetation, wetland and bare ground, together with the traditional farming practices, provides suitable habitat for some species not usually found on farmland elsewhere in Scotland.

Woodland

Natural woodlands contain trees and shrubs of several kinds and varying ages, which provide both a diverse food resource and shelter at different heights and densities, supporting the most varied populations of woodland birds, and summer visiting species. Clearings and rides give space for song flights and clear paths for hunting; dense shrubby cover is important for garden warblers and blackcaps; thickets close to the wood's edge for whitethroats; and a combination of tall trees and rough ground vegetation attract chiffchaffs. Most woods have natural holes that serve as nesting sites.

Two particularly Scottish birds are closely associated with the native pinewoods that now survive in only a few areas – the very local crested tit, and the more widespread crossbill. Today, non-native conifers make up a very large proportion of Scotland's woodland. The bird populations of these woods change with their stage of growth from common woodland species in the early stages to roosting birds, such as woodpigeons, when trees are mature.

Watching woodland birds is seldom easy, as many of them are hidden amongst the vegetation or high in the treetops. Recognition of songs and calls is therefore especially useful. Early morning is the best time to hear bird song – some species are active earlier than others, so the dawn chorus involves a gradually increasing selection of songs.

Freshwater

Some birds spend virtually the whole of their lives on the water, coming onto land only to nest. Others use lochs and reservoirs as safe roosting and bathing places. Some simply take advantage of the many insects on the water's surface.

Shallow lowland lochs usually support a good growth of plants, as well as eels and other fish, attracting swans, ducks, grebes and coots. Deep, cold lochs often have good populations of fish, making fishing ospreys frequent visitors, while Highland lochs with stony shores attract sandpipers and common gulls in summer. In slow-flowing lowland rivers fish can be readily spotted from above and sandy banks provide ideal breeding habitats. Along fast-flowing upland streams the insect larvae attract dippers and grey wagtails. Reedbeds or marshy ground adjoining low-lying lochs and rivers provide safe havens.

Water birds are most active early and late in the day, and often have a 'siesta' in the afternoon. In midsummer there is a period when many of the ducks 'disappear' while they are moulting; most dabbling ducks simply stay hidden. At any time of year it is worth looking along the shoreline for birds that are resting on land. Identifying water birds is easiest when the light is good and the water surface calm, but even in less favourable conditions, many species can be recognised by their shape and behaviour.

Heath and Hill

A large proportion of Scotland's countryside can be described as 'heaths', which include golf courses and other gorse-dotted rough ground as well as heather moorland and 'hills', ranging from the grassy hills in the Southern Uplands to the bare granite of the Cairngorms and the scree-sloped mountains of the northwest.

Buzzards, kestrels, hen harriers and short-eared owls sometimes hunt over rough grasslands and heaths, also populated by curlews and skylarks, with wheatears near stone dykes or rock. Heathery moors are important breeding grounds for birds such as merlins and golden plovers. Where the moors are broken up by rocky gullies, ring ouzels are often present, as well as wrens. In the far north the bare and exposed moors near sea level have their own distinct breeding populations. Cliffs and glens among the higher hills provide nesting sites for golden eagles, peregrines and ravens, all of which hunt – or scavenge – over wide areas.

Many of the heath and hill birds can be seen on ground to which the public has unrestricted access. Elsewhere it is advisable to stay on right-of-way tracks or obtain permission, especially from August until December when shooting may be in progress.

Coast

In summer, rocky coasts offer some of Scotland's most thrilling birdwatching. The best thing to do is to sit on the cliff top and look across a narrow inlet at the birds nesting opposite on rock shelves or ledges. Rocky islets, low headlands or sea-washed rocks attract varied birdlife throughout the seasons.

Sand and shingle beaches and dunes are used as resting areas but also provide nesting sites for some birds, such as ringed plovers and oystercatchers. During migration periods waders of several species may also spend time there. Late summer and spring are the best times for seeing the sanderling. The machair, a mosaic of shallow pools, flower-rich grassland and cropped ground, backs many Hebridean beaches and is an especially important habitat for breeding waders.

Muddy estuarine shores are of great importance to birds from autumn through to spring. For rewarding birdwatching at an estuary go shortly before or after high tide, when the birds are either being forced to come closer as the tide rises or are following it out as it falls. On some estuaries as many as 20 different waders occur annually. Many of Scotland's seabirds can also be watched during ferry trips between islands. Guillemots, razorbills, gannets, kittiwakes and puffins can often be seen in summer near the Hebrides and Northern Isles.

RED-THROATED DIVER

Gavia stellata
Learga Mhòr

Identification
The slender, up-tilted bill is held slightly raised.

Habitat Nests beside small moorland lochs and flies to the sea to feed.

Voice A goose-like cackle in flight; its display involves high wailing calls, often in duet.

When & where Mainly in the Northern and Western Isles and northwest Highlands. From September to March large numbers are off the east coast.

BLACK-THROATED DIVER

Identification The bill is straighter, heavier and held more level than the red-throat's.

Gavia arctica
Learga Dhubh

Habitat Large lochs, often with islands, provide both nesting sites and fishing grounds.

When & where Breeds in the northwest Highlands and western islands. Small groups winter around the Hebrides and Northern Isles and off the east coast.

LITTLE GREBE (or DABCHICK)

Tachybaptus ruficollis
Spàg-ri-Tòn

Identification
This is the smallest grebe, with a distinctive stocky appearance due to its short neck and blunt, rounded tail-end. In summer its brownish colour is relieved by chestnut cheeks and throat and a pale yellowish spot at the base of the bill. They fly more readily than the other small grebes and often dive with a noticeable splash.

Habitat Lochs and ponds with plenty of cover.

Voice Little grebes give a loud, whinnying trill in the breeding season, often calling from cover.

When & where Breed mainly in the lowlands. Some winter in sheltered coastal waters.

GREAT CRESTED GREBE

Podiceps cristatus
Gobhlachan

Identification A short black crest and black-tipped chestnut ruff around the top of its long neck distinguish this, the largest, grebe in summer. In winter its crest is barely detectable and its ruff disappears, leaving it very white-faced. Its silky white breast looks brilliant when it tips sideways to preen. They often float with necks drawn back and bills tucked in.

Habitat Low-lying lochs with enough emergent vegetation to provide cover for their nests and a good supply of eels and small fish.

Breeding They perform elaborate displays, including head-shaking with raised crests and a 'penguin dance' when the pair, each carrying weed, rise upright on the water, face to face.

Voice Harsh and guttural growls, croaks and barks.

When & where Breed mainly in the central lowlands. In late summer many move to the coast, where they winter in sheltered waters. Return to the breeding lochs may take place as early as February in winter.

SLAVONIAN GREBE

Podiceps auritus
Gobhlachan Or-Chluasach

Identification In summer, this species can be distinguished from the other small grebes by the conspicuous golden 'shaving brush' extending from in front of the eye to the back of the head, where it forms short 'horns', and by its chestnut neck, upper breast and flanks. In winter, when small parties may be seen on inshore coastal waters, it looks more cleanly patterned than little (p.17) or black-necked (p.20), with its black crown sharply separated at eye level from white cheeks, and its neck white except for a dark line down the back.

When & where During the breeding season, April to October, it occurs locally on highland lochs with some emergent vegetation, mainly in Inverness-shire.

BLACK-NECKED GREBE

Podiceps nigricollis
Gobhlachan na
h-Amhaiche Duibhe

Identification The black neck and upper breast, steep black forehead and high crown and drooping fan of golden feathers behind the eye are distinctive in summer, when this grebe is present on a few rich low-lying lochs in the central lowlands. Outside the breeding season the black-necked looks dingier than a Slavonian grebe, with cheeks and neck smudged greyish, while its longer body and fine up-tilted bill distinguish it from the little grebe (p.17).

When & where Small numbers winter around the coast, especially in Loch Ryan.

FULMAR

Fulmarus glacialis
Fulmair

Identification Fulmars glide on stiffly straight wings, using their large webbed feet as rudders, and often hang motionless on upcurrents along the cliff face. They fly low over the sea, tilting first one way and then the other, flapping only occasionally. They have 'tubenoses' – heavy-looking beaks with an obvious nasal tube – and no black on the wings. Both adults and chicks sit back 'on their haunches' and never stand up straight-legged like a gull.

Feeding Fulmars float buoyantly on the ocean surface as they forage for small marine creatures and offal discarded by fishing boats.

Breeding They nest separately, usually on a ledge overhanging the sea, on cliffs all round the coast and also in a few places inland.

Voice They cackle and head-wave at the nest but away from the cliffs are generally silent.

When & where Fulmars are entirely absent from their breeding cliffs only from mid-September until November.

STORM PETREL

Hydrobates pelagicus
Luaireag

Identification This tiny seabird is most likely to be seen from a boat, as it flits low over the waves, its feet often pattering the surface of the water.

Breeding Storm petrels breed on islands in the north and west and change over at their musty-smelling nesting burrows at night.

Voice They have a peculiar purring call ending in a hiccup.

When & where Found on many offshore islands. On Mousa, Shetland the birds nest in the dry stone walls of the ancient broch. They winter far at sea.

MANX SHEARWATER

Puffinus puffinus
Sgrab

Identification Manx shearwaters are most likely to be seen at sea, where they look alternately black and white as they glide low over the waves, tipping first one way and then the other and only rarely flapping.

Voice As they fly to their burrows at night they give weird screams and cackles.

When & where They breed in a large colony on Rum and a few smaller ones on islands in the west and north, and on summer evenings gather in rafts off these sites. In autumn parties can be seen all round the coast, as they leave Scottish waters to winter off the coast of South America.

GANNET

Morus bassanus
Sùlaire

Identification Goose-sized and conspicuously white and black, adult gannets are more distinctive than immature birds, which have variable amounts of brown on their back and wings. Even in areas where they do not breed gannets can often be seen offshore, flying in 'strings' or circling and plunge-diving from a height with folded wings.

Habitat They nest on islands, mainly in the far north and west, but also in large colonies on the Bass Rock and Ailsa Craig.

When & where The nesting colonies are occupied from March to August. Gannets are scarce in Scottish waters between December and February.

SHAG

Phalacrocorax aristotelis
Sgarbh an Sgùmain

Identification Their plumage has a green/bronze gloss and their short crests show best when blowing in the wind. The dark-brown young birds lack the whitish bellies of young cormorants. When swimming shags hold their heads horizontal, not uptilted. Like cormorants they often stand with wings spread, but seldom perch on anything other than rocks.

Voice Shags at the nest often hiss with wide-open bills, exposing their bright yellow mouths.

When & where Shags are more widespread and more maritime than cormorants. They breed on all exposed rocky coasts, usually nesting on low ledges or in caves.

CORMORANT

Phalacrocorax carbo
Sgarbh

Identification Its white lower face distinguishes this species from the smaller shag, and it frequents sheltered coastal waters rather than sites exposed to rough seas. Cormorants swim low in the water with head and bill tilted upwards; they dip their heads below the surface and dive frequently. They readily perch on buoys and posts, sitting upright and often with wings 'hung out to dry'.

When & where Most nest on rocky islands but in winter some move inland, to lochs and rivers, where they regularly perch on trees.

GREY HERON

Ardea cinerea
Corra-Ghritheach

Identification The heron's long neck and legs are distinctive when extended, but it looks quite different when standing with its head hunched between its shoulders. In flight its head is always drawn back, so that it appears to have a short thick neck, and its legs project beyond its tail.

Habitat In many areas herons are associated with freshwater, feeding in marshes and shallow lochs or on damp farmland, but on the west coast they are found along the seashore.

Feeding They feed solitarily but nest in small colonies, usually in woodland or conifer plantations.

Voice A harsh 'kraaak' call is often given in flight, while nesting birds produce a variety of bill-snapping, croaking and other noises.

When & where The tidal-feeding west-coast population is unaffected by severe winters, whereas in the colder east numbers decline.

MUTE SWAN

Cygnus olor
Eala

Identification The black knob at the base of the mute swan's orange bill, its curved neck and noisy wing beats distinguish it from the whooper swan.

Feeding Feed on aquatic plants, often using their feet to help them remain up-ended.

Breeding They breed solitarily on lowland freshwater and also by sheltered tidal waters.

When & where In late summer moulting flocks gather on large lochs and estuaries, mainly in the east.

WHOOPER SWAN

Cygnus cygnus
Eala Bhàn

Identification Straighter neck than a mute swan, with no bill knob. The black and yellow bill pattern varies, but the tip is always black.

Voice They are gregarious and noisy, giving their bugling 'whoop-whoop', on the ground and in flight.

When & where They winter mainly in lowland farming areas, where they roost on lochs and feed over stubble and potato fields as well as in lochs and marshes.

WHITE-FRONTED GOOSE

Anser albifrons
Gèadh-bhlàr

Identification This species is distinguishable, at all ages, by its orange legs and smallish orange-yellow bill. Young birds lack the adults' white forehead and black bars across the belly. In flight head, neck, back and wings look uniform in colour.

Habitat They feed over rough, boggy ground as well as farmland and often roost on moorland lochs.

When & where They winter in the west, such as on Islay and the Kintyre peninsula, and in the southwest. Flocks are usually small, seldom more than one hundred birds.

PINK-FOOTED GOOSE

Anser brachyrhynchus
Gèadh

Identification Pink-footed geese can be distinguished from the larger greylag by the contrast between their markedly dark-brown head and neck and grey-brown body, and by their small, mainly pink, bills.

Habitat They mainly roost on lochs and estuaries, less often on moorland. Most of their roosts are long-established.

Feeding They often feed and roost in very large flocks and are warier than greylags. They start to arrive from Iceland during the second half of September, feeding over rich farmland. On arrival they glean grain, especially barley, from stubbles, moving onto cleared potato fields and grassland in autumn, and towards spring they feed mainly on grass but sometimes on young cereal crops.

Voice Their usual call, often heard when the birds are in flight, is a high, occasionally squeaky, two- to three-syllable 'pink-pink'.

When & where They are most numerous in lowland east and central Scotland.

GREYLAG GOOSE

Anser anser
Gèadh-Glas

Identification A uniformly grey-brown goose, with pink legs and a heavy bright orange bill, this is the ancestor of the farmyard goose. In flight it shows silvery-grey 'shoulder' triangles which contrast with the dark hind edge and tips of the wing.

Voice Its voice is much deeper than a pinkfoot's, a trisyllabic 'ang-ung-ung', rather like a farmyard goose.

Habitat They sometimes mix with pinkfeet when feeding but usually roost separately. They tend to be in smaller flocks and use a wider selection of roosts, sometimes spending the night on a river bank. Immigrants winter in much the same areas as the pink-footed geese, but are more widely scattered. They occasionally damage turnip crops in hard winters, when other food is difficult to obtain.

When & where Greylags from Iceland arrive from early October onwards. This species also breeds in Scotland, as a native resident in the Western Isles and northwest Highlands. They remain in or near their breeding area throughout the year.

BARNACLE GOOSE

Branta leucopsis
Cathan

Identification A pure white face and forehead and jet-black neck at once distinguish the barnacle goose. Its back is grey barred with black and its underparts are silvery grey but the general effect is of a smart black and white bird. Its legs and small bill are black. From below the black neck and upper breast contrasts with the whitish belly. This species is very gregarious, feeding and flighting in tight packs, and is noisy and quarrelsome.

Feeding Barnacle geese feed on coastal saltmarsh (merse) as well as on farmland.

When & where Wintering barnacle geese are less widely distributed than the grey geese and occur mainly in two distinct areas. Those from Spitzbergen frequent the shores of the Solway while those breeding in Greenland visit islands off the north and west coasts, with by far the biggest concentration on Islay. Elsewhere in Scotland barnacles occur on passage or as stragglers among grey geese. The Solway birds start to arrive in late September and the Islay flock in the second half of October; both leave again in April.

CANADA GOOSE

Branta canadensis
Gèadh Chanada

Identification This large, dark goose was introduced to Scotland many years ago, initially on ornamental and park ponds. It has now become established in widely scattered areas and the population is steadily increasing.

When & where Canada geese in this country are not migratory, but from June to August large numbers, many of them from Yorkshire, gather to moult on the Beauly Firth.

SHELDUCK

Tadorna tadorna
Cràdh-Ghèadh

Identification This large and distinctively coloured duck looks black and white at a distance. Young shelduck are brownish above and white below.

Feeding It is more often seen feeding over wet mud or sand, with a scything motion of the head, or standing around near its nesting burrow, than on the water.

Breeding Shelduck breed solitarily beside sandy coasts and estuaries, with a very few nesting inland. They are territorial and have frequent noisy encounters with neighbouring pairs.

When & where
In August–October most leave the breeding areas to moult, many going to the inner Forth.

MALLARD

Anas platyrhynchos
Tunnag Fhiadhaich

Identification Apart from the drake's glossy green head, the best distinguishing feature of a mallard is its broad navy/purple speculum narrowly edged with white, which is conspicuous in flight and usually visible in the closed wing. When the birds up-end to feed, or stand on land, their bright orange legs and feet are very obvious. The females give the familiar 'quack'. They are adaptable and often tame, taking a wide variety of foods and happily living close to man.

Breeding Their breeding season starts early and from February onwards they can be seen chasing around in their fast display flights. When moulting their flight feathers, males in June–August and females a few weeks later, they are often difficult to spot as they spend much of their time amongst reeds.

When & where Mallard breed almost anywhere other than the higher hills and feed on lochs, mud flats and farmland. In winter they desert upland areas and join large flocks at estuaries and lowland lochs.

TEAL

Anas crecca
Crann-Lach

Identification This small, neat duck spends much of its time in cover. At a distance the drake looks grey with a dark head; in flight the metallic green speculum, bordered with white, shows up well in both sexes. Teal rise almost vertically off the water and fly very fast and erratically.

Breeding For breeding, teal need a combination of open water and marsh or damp scrub vegetation; they nest in wetlands from low ground well up into the hills.

Voice The male has a musical and carrying 'krik' call and the female a high 'quack'.

When & where In winter, flocks of several hundred gather on the Solway and some east-coast estuaries but most remain, thinly scattered, on lochs and marshes. They do not feed over fields as mallard do.

SHOVELER

Anas clypeata
Gob-Leathann

Identification They seldom leave the water and when not feeding usually hold their heads 'hunched between their shoulders'. The long head and distinctive shovel-shaped bill give this species a rather unbalanced appearance and make it look in flight as though its wings are set far back.  The male's breeding plumage is unmistakable, and both sexes show pale blue shoulder patches when flying.

Feeding Shovelers feed in a characteristic manner, swimming with head low and neck stretched forward and sweeping their bills from side to side on the water surface. The bill is specially adapted to filter out tiny particles of food as the water passes through it.

Breeding Small numbers of shoveler breed regularly on rich lowland lochs in the southwest, the central lowlands, Orkney and the Outer Hebrides, and occasionally elsewhere.

When & where Most widely scattered and abundant in September–October, when Scottish breeding birds move south and immigrants arrive; few winter here, however, and most leave before the first hard frosts.

PINTAIL

Anas acuta
Lach Stiùireach

Identification The drake's white neck and long tail make it easy to recognise. Female pintail have longer necks than mallard (p.35) or wigeon (p.39), with which they are most likely to to be confused; they are slimmer than mallard and have grey bills and more pointed tails. Pintail fly fast, with quicker wingbeats than mallard, and usually feed by up-ending.

When & where Flocks of pintail occur regularly only on the Solway, Moray, Cromarty and Dornoch Firths, but single birds or small parties occasionally appear elsewhere on the coast and on lochs inland.

Swans, Geese and Ducks

WIGEON

Gavia stellata
Gael: Learga Mhòr

Identification Male wigeon can be distinguished in flight by their conspicuous white forewing; females, and males not in their distinctive breeding plumage, are a much redder brown than the larger mallard (p.35) and have proportionately smaller, blue bills. Wigeon rise straight from the water and fly fast and direct.

Habitat Wigeon breed in small numbers near pools and lochs on moorland. In winter many more are present and are more widely distributed, on lochs and estuaries and also on grassland along river banks.

Feeding Feed by up-ending but also regularly graze on short grass and on eelgrass growing on estuarine mud. When on freshwater they often scavenge plant material brought up by coots and swans.

Voice Drakes give a whistling, musical 'whee-oo' and ducks a purring growl.

When & where Found mainly in the central and northern Highlands, but also locally in southern Scotland and a few islands. In winter, wigeon are present in thousands on east-coast estuaries.

POCHARD

Aythya ferina
Lach Mhàsach

Identification The drake pochard's chestnut head, grey back, and black breast and tail are distinctive; female goosanders and red-breasted mergansers (p.44–5) also have chestnut heads but lack the black patches and are a different shape. Pochard ducks can be confused with female tufted ducks (p.41), with which they often associate, but are paler brown and have less rounded heads, with a longer more sloping forehead; the sides of the head are usually noticeably lighter than the crown. In flight both sexes show a rather indistinct greyish wing bar, whereas the tufted duck's is vivid white. Pochard spend much of the daytime floating around, apparently asleep, and dive most actively at dawn and dusk.

When & where This species is present from late summer to spring on many low-lying lochs and reservoirs, most often in small parties but with flocks of hundreds sometimes present, and even thousands on Loch Harray in Orkney. The first to arrive in this country are usually all drakes, the females appearing a week or two later. Very small numbers breed on a few shallow lochs which have abundant aquatic plant life and plenty of cover.

TUFTED DUCK

Gavia stellata
Gael: Learga Mhòr

Identification This is the most common and widely distributed of the resident diving ducks. Drakes in breeding plumage have smart black and white plumage and a drooping crest, but females and moulting males are a nondescript dark brown, with only a faint crest showing at the back of the round head. Some females have a narrow white band round the base of the bill similar to, but smaller than, that of a female scaup (p.42). At close range the yellow eye can be seen and the long white wing bar is very obvious in flight. This species is usually silent, but its wings make a whistling sound as it flies.

Habitat Tufted ducks are most often seen on still freshwaters, sometimes occupying quite small lochs during the breeding season but moving to larger ones in winter, and occasionally resorting to large rivers or estuaries in hard weather. They are most abundant in the lowlands, and regularly visit park ponds.

When & where The numbers on lochs are often quite small, but even when large numbers gather in winter they do not form a close flock.

SCAUP

Aythya marila
Lach Mhara

Identification Males can be distinguished from drake tufted ducks (p.41) by their grey backs. Females can be confused with female tufted ducks, which sometimes also have a white band around the base of the bill, though this is usually broader and more conspicuous in the scaup.

When & where Scaup are usually seen at the coast, where they winter in large flocks at a few sheltered sites, including the Forth near Leven, the Solway between Carsethorne and Southerness, Loch Ryan, the Clyde, Loch Indaal on Islay, and the Dornoch Firth.

GOLDENEYE

Bucephala clangula
Lach a' Chinn Uaine

Identification Goldeneye have small bills and almost triangular heads, high at the crown. Males can be distinguished, in all plumages, from drake tufted duck by their white chests and females by the combination of chocolate-brown head, white collar and greyish body, with a white wing patch towards the tail. In flight both sexes show conspicuous square white patches reaching almost to the front of the wing. Goldeneye are restless, active and agile, diving frequently and able to rise more rapidly from the water than other diving ducks. They often dive all at the same time, so that a whole group suddenly vanishes from sight.

Breeding Breeding birds are usually found on small lochs surrounded by woodland; they nest in holes and on Speyside many occupy nest boxes.

When & where There is a small but slowly expanding breeding colony on Speyside, and occasionally birds summer in other areas, but most goldeneye are winter visitors. Wintering birds are widely though thinly scattered over lochs and rivers, but gather in bigger concentrations on the large estuaries near sewage outfalls.

GOOSANDER

Mergus merganser
Lach Fhiacailleach

Identification The creamy, almost pinky, white of the breast and sides of a drake goosander in breeding plumage stands out against the glossy dark-green head. Females can be distinguished from red-breasted mergansers by the clear-cut junction between chestnut and grey-white on the neck. In this species the crest droops down the back of the head and is often not obvious. Goosanders frequently swim with head submerged before diving for fish; family parties often all dive at the same time.

Habitat Much more a bird of freshwater than the merganser, it breeds on large lochs and fast rivers.

Voice Goosanders and mergansers are usually silent.

When & where Breeds mainly in the central highlands and southwest. In winter it is more widely distributed, occurring on many lochs not occupied in the breeding season, especially in the south and east. From May to September moulting goosanders gather on both fresh and tidal waters in a few areas. Large wintering flocks are on the Tay and Beauly Firths.

RED-BREASTED MERGANSER

Mergus serrator
Síolta Dhearg

Identification Markedly long and slim bodied and billed, this fish-eating duck can be confused with the similar but larger goosander. Males in breeding plumage are the easiest to distinguish: they are dark chestnut on the 'chest' where a goosander is white, and grey, not white, along the sides of the body. The most reliable distinguishing features in females are the red-breasted merganser's ragged wispy crest, which sticks up and out at the back of the head, and the gradual merging of chestnut with grey on its neck. When swimming, mergansers often jerk their heads forward in time with each push of their feet. Females of the two species are difficult to distinguish in flight.

Habitat Mergansers breed on freshwater, sea lochs and rivers.

When & where Breed most abundantly in the northern and western highlands and the islands. After breeding they move to the coast, where flocks gather to moult, before dispersing to winter on tidal water. Large numbers of immigrants are also present in the Firths of Forth, Tay and Moray in winter.

COMMON SCOTER

Melanitta nigra
Tunnag Dhubh

Identification Often visible only as a tight raft of all-dark birds well offshore. All members of a flock frequently dive at the same time, so that the whole raft vanishes from sight.

When & where This sea duck occurs most regularly off the east coast, in both summer and winter. Moulting flocks are present in July–August off Dornoch, north of Aberdeen, near Montrose, and in the Forth off Gullane; the main wintering flocks are in the Moray Firth, St Andrews Bay, and the Forth.

VELVET SCOTER

Melanitta fusca
Lach Dhubh

Identification Velvet scoters often accompany common scoters, from which they can readily be distinguished in flight, or when flapping their wings, by their white wing bar. On the sea at a distance it is difficult to identify them.

LONG-TAILED DUCK

Clangula hyemalis
Lach Bhinn

Identification Long-tailed ducks are active and agile surface divers, whose small size, round head, small bill and piebald pattern help to distinguish them from other sea ducks.

Voice In spring the males give a musical yodelling call, often in chorus.

When & where Winter in large numbers in the Moray and Dornoch Firths, the Northern Isles and off the east coast.

EIDER

Somateria mollissima
Colc

Identification Moulting and young eider drakes are often patchy dark brown and white, but can always be identified by the long even slope from bill to crown. The female is the only large uniformly mottled brown sea duck. Take-off from the water is laboured and they usually fly low and in single file.

Feeding Eiders both surface-dive and up-end in the shallows to feed largely on mussels, leaving crushed shells in their droppings.

Breeding Eiders breed on all open coasts, most often in rocky areas but also locally among dunes. Courting males croon 'ahooo' in chorus and perform an elaborate head-waving display. Incubating females are very tame, but are so well camouflaged that they are not easy to see. A crèche system is operated, with a few females looking after several broods of sooty brown chicks, which sometimes have to walk a long way, or descend cliffs up to 100m high, to reach the sea.

When & where In July–September moulting flocks gather off the coast, the largest at the mouth of the Tay.

SPARROWHAWK

Accipiter nisus
Speireag

Identification Sparrowhawks have short rounded wings, long barred tails, heavily barred underparts, and long yellow legs; in flight they can sometimes be confused with cuckoos (p.105). The female is much larger than the male, brown rather than grey on the back, and whitish not red-brown below.

Habitat Sparrowhawks occur wherever there are woodlands thick enough to provide good cover but open enough to allow easy flight. They often visit gardens and may perch on fences or rooftops.

Feeding They hunt by flying low and fast along hedges and among trees, and grabbing unsuspecting small birds with their feet.

Breeding In spring, they soar high above tree level and circle in display. The population is probably now larger than ever before.

Voice They are usually silent but when intruders are near a nest they give loud, rapid 'kek-kek-kek' calls.

When & where This species occurs on the mainland and some of the inner islands.

HEN HARRIER

Circus cyaneus
Clamhan nan Cearc

Identification Harriers are slim-bodied birds which fly low and buoyantly. At a distance the male looks almost like a herring gull (p.91) in flight, but has rounder wing-tips, a white rump and a grey tail. The larger female's white rump contrasts more conspicuously with the streaky-brown of back and tail. Young birds of both sexes resemble females.

Habitat A bird of heather moorland and very young conifer plantations. In winter, small parties roost communally in marshy areas and hunt over lower ground.

Feeding When quartering the ground in search of small rodents and birds, hen harriers frequently glide with wings held in a shallow 'V'. Their display flight involves switchbacks, rolls and the aerial passing of food from male to female.

Voice Adults give a high chattering 'kek-kek-kek' near the nest and often dive at intruders.

When & where The hen harrier breeds in scattered areas of the mainland and some of the western islands.

BUZZARD

Buteo buteo
Clamhan

Identification Buzzards look 'neckless' when soaring, with wings held in a shallow 'V' and rounded tail widely spread; in direct flight shallow wing beats alternate with short glides. When perched they sit upright and have a bulky appearance; their bright yellow lower legs are usually visible. They vary considerably in colour, some birds being much paler or darker than the usual darkish brown.

Feeding Buzzards hunt by pouncing, from a low hover or a perch, on rabbits and other small animals, as well as carrion. In the past they were thought to be a threat to game birds, and despite legal protection are still shot or poisoned in some areas.

Voice Buzzards call frequently, with a plaintive far-carrying 'meeooo' which is distinctive.

When & where This is the commonest large bird of prey in areas where there is a mixture of rough open ground and scattered woodland, especially on fairly low hilly ground.

GOLDEN EAGLE

Aquila chrysaetos
Iolaire

Identification Although size is not easy to gauge at a distance, the eagle's flight silhouette is different from that of the much smaller buzzard: the wings are longer and less rounded, the head projects further, and the tail is longer and less spread. An eagle hunting along a hillside flies low, with deep and powerful wing beats, and strikes its prey with its claws while still moving fast.

Habitat They hunt over open upland country ranging from heather moorland to mountain tops.

Breeding They are very early nesters and pairs can be seen soaring in display flight on fine days in mid-winter. Eagles are sensitive to disturbance so nesting sites should not be approached in February–March, when eggs can quickly become chilled.

Voice Eagles are usually silent.

When & where
Golden eagles are widely scattered over the highlands and larger western islands.

WHITE-TAILED EAGLE

Haliaeetus albicilla
Iolair-Mhara

Identification This huge bird has been described as looking like a flying door! It has a wedge-shaped tail and unfeathered lower legs, whereas the golden eagle's (p.52) tail is square-ended and its legs are feathered.

When & where White-tailed eagles were exterminated by persecution early last century but were recently reintroduced on Rum. Several pairs now breed on the west coast north of the Clyde and wandering birds are occasionally seen in other areas, sometimes inland.

OSPREY

Pandion haliaetus
Iolair-lasgaich

Identification In flight the dark-above, white-below osprey looks a bit like a black-backed gull (p.89), but its wings are longer and less pointed and have a pronounced angle at the 'wrist'. Seen from below it has a similar pattern to a buzzard (p.51), but longer and narrower wings.

Habitat Ospreys occupy a territory which includes large trees, often pines, for nesting, and good fishing in lochs or rivers. The same nests are used year after year, with fresh sticks and lining added annually.

Feeding Ospreys fish most actively early in the day, splashing feet-first into the water and rising with their catch firmly gripped in their talons.

Voice Their call, usually given near the nest, is a short whistling 'pew'.

When & where Most ospreys arrive from their African wintering grounds in early April. Breeding ospreys are now quite widely scattered from the central lowlands northwards.

KESTREL

Falco tinnunculus
Clamhan Ruadh

Identification The kestrel is the only small bird of prey which spends much of its time hovering, with flickering wings and depressed tail; on sighting prey it drops suddenly and quickly to seize it on the ground. Kestrels often perch in an upright position on telegraph poles and dead trees. Heavy barring with black makes the female's tail, back and wings look much darker than the male's. In direct flight the wing beats are fast and shallow, the tail looks long and narrow and the wings long and pointed. Kestrels frequently soar, with tail fanned.

Habitat Resident wherever there are suitable nesting sites, e.g. buildings or quarries with ledges, large tree holes, or old crow nests. They generally hunt over open ground, such as farmland and verges, less often in open woodland or over waste ground.

Voice Usually silent, but near the breeding site give a shrill high 'kee-kee-kee' call.

When & where They are only scarce in the Northern Isles and Outer Hebrides.

PEREGRINE

Falco peregrinus
Seabhag

Identification In normal flight, peregrines alternate fast wingbeats and glides, and sometimes hover. Crow-sized, it is compactly built and has sharply-pointed wings and a tail that tapers towards the tip. Juvenile birds are browner than adults and streaked below, instead of barred. It is a cliff-nester.

Voice A harsh, screaming chatter when disturbed.

When & where It is widespread in mountainous areas and breeds occasionally on the coast.

MERLIN

Falco columbarius
Mèirneal

Identification This small falcon flies fast and low, with rapid turns and climbs. Its pointed wings distinguish it from the sparrowhawk (p.49) and its colouring and flight from the kestrel (p.55).

When & where In summer, merlins are thinly scattered on moorland from Shetland to Galloway.

PHEASANT

Phasianus colchicus
An Easag

Identification Their very long tails make males easy to identify, even though their colours vary. Pheasants often seem reluctant to take wing and will run to and fro beside a fence rather than fly over it. When they do decide to fly the take-off is noisy, and they usually go only a short distance before seeking cover.

Habitat Pheasants are often seen feeding in woods or fields, or walking along roadsides. They typically feed in a crouched position, frequently raising their heads to check for danger.

Voice After giving their strident crowing 'korrk-kok' call the males may whirr their wings briefly.

When & where Essentially a lowland bird, pheasants are scarce in the central and northwest Highlands and the islands. Many are artificially reared and released for shooting, and large numbers are often present in autumn close to rearing and feeding sites.

CAPERCAILLIE

Tetrao urogallus
Capall-Coille

Identification Turkey-sized capercaillies are most likely to be seen in flight, as they crash away through the trees after being flushed; much of the noise is from their wing beats – they are surprising agile when flying. The huge dark males are unmistakable, but females might be confused with female red or black grouse (p.59-60); size and the capercaillie's reddish breast and broad fan tail are the main distinguishing features. Like black grouse they indulge in communal display. Some males are very aggressive during the breeding season and will attack people or vehicles.

Habitat Capercaillies breed in conifer woods, most often of mature pine, but sometimes feed on adjoining fields or among deciduous trees.

Voice The male's 'song' is a strange series of accelerating clicks, ending with a cork-pulling 'pop'.

When & where They are found mainly in the eastern Highlands, with smaller numbers as far west as the Loch Lomond islands.

BLACK GROUSE

Tetrao tetrix
Coileach-Dubh, Liath-Chearc

Identification Black grouse are most obvious in spring, when gathered at their breeding 'leks', which are usually on open ground near woodland. There the males display with wings lowered and tails raised, to show the white underside, to a chorus of bubbling musical 'roo-koo'ings', which carry a long way. The females stand around nearby. In winter black grouse often sit in trees.

When & where They are widespread in the Southern Uplands and much of the Highlands but scarce north of the Great Glen and absent from most islands.

RED GROUSE

Lagopus lagopus
Coileach-Fraoich,
Cearc-Fhraoich

Identification Its red comb and angry breeding season 'go-bak, go-bak' calls make the male easy to identify, especially when it stands on a prominent rock. The well-camouflaged females are less obvious, as they tend to skulk in the heather. When flushed, red grouse take off with an explosive burst of whirring wing beats, then glide with down-curved wings.

When & where Found mainly on heather moors but sometimes visit nearby farmland in winter. They are widespread in suitable habitat, but have decreased in some areas due to moorland afforestation.

PTARMIGAN

Gavia stellata
Gael: Learga Mhòr

Identification Its white wings identify the ptarmigan at all seasons. In winter both sexes are almost completely white; in summer they are mottled brownish, with varying amounts of white below, and can be difficult to spot as they crouch and 'vanish' against a similarly mottled background. When flushed they give a rattling croak.

When & where Ptarmigan breed on mountain heaths, mainly above 800m in the central highlands but at lower levels in the northwest. They can often be seen near the top of the Cairngorm and Glenshee chairlifts.

GREY PARTRIDGE

Gavia stellata
Gael: Learga Mhòr

Identification Its orange face and throat immediately identify this dumpy bird, whose short neck and drooping tail give it a rather sad hunched appearance. In flight a dark inverted horseshoe mark on the pale belly can be seen. Young birds lack colour on face and underparts and are streaky brownish all over. When disturbed they crouch then run for cover; if forced into flight they usually stay close together and fly low. Family parties stay together, as 'coveys', well into the winter, and although wary are easy to watch from a car. They feed most actively early and late in the day.

Habitat Grey partridges feed over the fields and nest among rough growth along the bottom of hedges. Hedge removal and spraying of crops has adversely affected nesting sites and the insects the partridge feed on.

Voice The calls of the grey partridge are a high creaky 'kree-vit' and a rapid cackled 'ek-ek-ek' as it takes flight.

When & where A bird of the lowlands.

RED-LEGGED PARTRIDGE

Alectoris rufa
Cearc-Thomain
Dhearg-Chasach

Identification A more strikingly coloured and patterned bird than the similarly shaped but smaller grey partridge. Young birds have only indistinct face markings.

Voice Calls are a harsh 'shrek' and a more conversational 'chuk-chuka'.

When & where An introduced game bird, this species is reared and released locally in lowland Scotland.

CORNCRAKE

Crex crex
Trèan ri Trèan

Identification They are shy birds, skulking among long vegetation, in meadows, marshy areas and weedy patches of arable fields.

Voice Corncrakes proclaim their presence night and day with a monotonous and distinctive rasping 'crek-crek'.

When & where This species now breeds regularly only in the western islands.

MOORHEN

Gallinula chloropus
Cearc-Uisge

Identification Its red bill and forehead, white line along the side and white patch under the cocked tail, distinguish the moorhen from other plump dark water birds. As it swims it nods its head and jerks its tail. It patters along the surface when taking flight and flies low only for short distances. Young birds are brown and do not have red bills.

Habitat They nest beside ditches and slow rivers as well as lochs and ponds.

Feeding Moorhens feed mainly at the water surface, only occasionally up-ending or diving, but also visit fields, especially in winter.

Breeding Moorhens operate an 'extended family' system, with young from the first brood helping to feed chicks from a later hatching.

Voice A liquid 'prruk' is the most distinctive note, but calls include chattering and squeaking sounds.

When & where Widespread in the lowlands, but scarce over much of the highlands and on most of the islands.

COOT

Fulica atra
Lach a'Bhlàir

Identification The white bill and forehead identify the bulky black coot, which has a tail-less look as its rounded back slopes down at the rear. Young birds are grey with whitish bellies. Their flight is laboured, with a long surface-pattering take-off and a breast-first splash landing. On land their gait is a waddling run, often with flapping wings. Coots are gregarious, noisy and aggressive; constantly indulging in territorial squabbles, with birds charging at each other with raised wings and lowered heads. They often form large flocks, especially in winter, when many immigrants are present.

Feeding Coots feed on underwater plants, diving for them with an audible 'plop', or from scavenging vegetation pulled up by swans or ducks.

Voice The commonest call is an explosive 'kut', from which the bird gets its name.

When & where Although they too sometimes forage over farmland, they are more dependent than moorhens upon still water and occur mainly on large lochs and reservoirs on low ground.

WATER RAIL

Rallus aquaticus
Snagan Allt

Identification Water rails are seldom visible, spending most of their time skulking in reedbeds and similar dense marshy vegetation.

Voice They are very audible, emitting, often at night, an astonishing range of squeals, grunts and screams reminiscent of pigs squabbling for food.

When & where In summer they are widely but thinly scattered wherever there are extensive areas of suitable wetland. In winter, when immigrants arrive, they occur also in ditches and are sometimes seen along river banks during hard frost.

OYSTERCATCHER

*Haematopus
ostralegus*
Gille Brighde

Identification The oystercatcher's black and white plumage and long orange-red bill make it difficult to confuse with any other species.

Feeding They feed by stabbing into mud or soil, or poking around in rock pools and among seaweed, and can open shellfish with their strong bills. When feeding they are usually well spaced out, but at high tide they gather in tight flocks to rest on islets, sand bars and occasionally fields.

Voice Oystercatchers are very noisy, and on territory keep up prolonged piping duets and repeated 'pik-pik-pik' calls well into the night.

When & where In winter, oystercatchers are scattered along all types of low coast, with main concentrations in larger estuaries. Birds move inland from February onwards, when their loud 'kleep' contact calls can often be heard as they fly overhead at night. In summer it is one of the most widely distributed of the waders, breeding along the coast and also inland in fields, near rivers and on loch shores.

RINGED PLOVER

Charadrius hiaticula
Bòdhag

Identification The ringed plover's pied head and breast and white neck band distinguish it from other common small waders. It is plump and compact and its behaviour helps with its identification: it looks alert and busy as it moves in short runs, stopping to tilt forward to pick up food, then running on again.

Breeding Most ringed plovers nest on, or near, sand or shingle beaches. Some move far inland to breed on loch shores and river shingles; especially large numbers breed on the 'machair' of the Outer Hebrides. When intruders approach a nest the birds try to lure them away, by feigning a broken wing or fanning their tails and squealing like a rodent. They keep up a continuous plaintive 'queep'ing while an intruder is present.

When & where In winter the majority of ringed plovers are on estuaries, but many remain on Hebridean beaches, with some on most sandy or muddy coasts.

Waders

GOLDEN PLOVER

Pluvialis apricaria
Feadag

Identification This species' liquid 'tlooee' calls and rippling trills are among the most typical sounds of heather moorland and bog in summer, drawing attention to golden plovers when their spangled plumage makes them difficult to spot. When a mixed party is disturbed their fast, direct flight and pointed wings make it easy to pick out the golden plovers. The black chest and belly are lost in winter, when there might be confusion with knot (p.72) which feed in tighter flocks and have pale rumps and wing bars.

When & where They breed wherever there are suitable habitats on mainland and islands and are inland from March to July. In late summer they move towards the coast, where they winter on farmland and mud flats, often resting with lapwings (p.71) on grass or potato fields.

GREY PLOVER

Pluvialis squatarola
Trìlleachan

Identification This species can most easily be distinguished from the more common golden plover (p.69) in flight, when its whitish rump and wing bar and black 'armpits' can be seen. On the ground it might be confused with a knot, but is taller, has a shorter bill, and occurs in small loose parties, not tight flocks.

When & where The grey plover is a rather scarce winter visitor, which frequents muddy estuaries and occurs most regularly on the Solway, Forth and Eden.

LAPWING

Vanellus vanellus
Curracag

Identification Lapwings are probably the most familiar of the waders, with their wispy crests, contrasting black and white plumage and loud nasal 'pee-wit' or 'pees-weep' calls. In flight their wings look very rounded, and their wing beats are noticeably noisy. In the breeding season they perform aerobatic display flights, involving steep ascents, sudden dives, slow 'butterfly' flapping and much rolling from side to side, when the birds look alternately light and dark.

Habitat Lapwings are found on farmland, nesting on rough grassland or arable fields and wintering mainly on low-lying coastal grassland rather than on mud flats.

Feeding They stop frequently, with head tilted forward, to pick up or probe for small invertebrates.

When & where They start to flock from mid-June onwards on stubble, short grassland and potato fields. As winter approaches Scottish birds move nearer the coast. By mid-March most native lapwings are back on their breeding grounds.

KNOT
Calidris canutus
Luatharan Gainmhich

Identification Knots are among the most gregarious of the waders and usually occur in tightly knit, large flocks. They fly in aerobatic clouds, looking alternately light and dark as they wheel and turn. They are noticeably dumpy and in early autumn and late spring there may be traces of their russet breeding plumage on back and underparts.

When & where The main winter concentrations of knot are on the Forth and Solway.

SANDERLING
Calidris alba
Luatharan Glas

Identification Usually seen running like a clockwork toy along the very edge of the retreating waves on sandy beaches, stopping occasionally to peck with head low and tail high. Often call with a liquid 'twik-twik' as they fly low over the sea, and their wings appear to flicker.

When & where Present all year, but numbers are highest when migrating flocks pass through.

COMMON SANDPIPER

Actitis hypoleucos
Luatharan

Waders

Identification Apart from a conspicuous white wing bar, it has no distinctive markings but its behaviour is characteristic. It flies low, with fluttering wing beats alternating with glides, wings held below body level; when standing, it constantly bobs its head and body and wags its tail. Unlike many other waders they are not at all gregarious, even in winter; although quite a number may be present on an estuary they remain scattered around, and do not feed or roost together in a flock.

Habitat This is the only wader closely associated with upland streams, stony rivers and loch shores.

Voice When flushed it gives a shrill, ringing 'swee-wee-wee' call as it flies above the water.

When & where Common sandpipers breed wherever there is suitable habitat, well spaced out along a river or loch shore. When they desert the nesting area in late summer they move first to mud or shingle shores on the coast and later leave Britain to winter in Africa.

PURPLE SANDPIPER

Calidris maritima
Cam-Ghlas

Identification This dumpy little wader is closely associated with low rocky shores where low tide reveals seaweed. Its dark back and upper breast help it to blend into its background, but it is so tame that its yellow legs can generally be seen without difficulty. Purple sandpipers are usually in small parties, often with turnstones, and twitter in chorus, sounding rather like swallows. They fly low over the water and land with a sudden flutter; in flight the white sides to the tail contrast with the very dark back.

When & where Widespread on rocky shores from August to April.

DUNLIN

Calidris alpina
Pollaran

Identification When feeding over estuaries and beaches, dunlins move forward together in flocks, with heads down as they probe in the mud. Their shoulder-hunched attitude and fine slightly down-curved bills help in identification. Flocks frequently perform aerial evolutions, 'flowing' to and fro in a tight cloud, sometimes close to a flock of bulkier knots (p.72). Dunlins in winter plumage lack conspicuous markings but in flight the narrow whitish wing bars and white sides to the rump usually show up well. Their summer plumage, with the black belly patch, is distinctive.

Breeding Dunlins winter on the coast and breed on moorland. They nest solitarily, at altitudes ranging from sea level in the islands to about 1000m in the Cairngorms. Nesting pairs attract attention with their shrill 'referee's whistle' calls, or the male's trilling song given in display flight.

When & where From August to March dunlins are widely scattered around the coast. Their most important breeding grounds are in the Northern Isles, Caithness and Sutherland and the Uists.

WOODCOCK

Scolopax rusticola
Coileach Coille

Identification Woodcock are beautifully camouflaged, their heavily marbled red-brown plumage blending with dead leaves and bracken. Their eyes are set high on the sides of the head, so that they are able to watch for predators overhead. When flushed, they fly away noisily, quickly dropping back into cover. It performs its 'roding' display flight at dawn and dusk in spring and early summer. This takes it low over the trees with slow, almost owl-like, wing beats; as it flies it gives a series of deep croaks followed by a shrill sneeze.

Habitat This long-billed wader is most likely to be seen in fairly open deciduous woodland or along the rides in conifer forests.

Feeding Woodcock feed and nest on the ground in damp woodland with good ground cover; they are wary and seldom seen feeding, but sitting birds sometimes allow a close approach.

When & where Woodcock are widespread on the mainland and many remain in woodlands all year.

SNIPE

Gallinago gallinago
Naosg

Identification Snipe are more likely to be seen in flight than on the ground, where their camouflage colouring and secretive, solitary behaviour make them difficult to spot. In spring and early summer their display flights, often performed at dusk, attract attention. These involve 'chip-per, chip-per' calls and oblique downward dives accompanied by a loud, almost buzzing, drumming noise, produced by the passage of air through the spread tail. When flushed snipe give a hoarse, rasping 'skaap' and fly off in a series of erratic zigzags, before dropping into cover again. The very long, straight bill is always obvious.

Habitat Snipe require soft, marshy ground to feed in, as they probe deep into mud. They breed on boggy moorland, in marshes and on damp rough grazing with clumps of coarse vegetation.

When & where In winter they become scarce inland, as frozen ground makes feeding impossible. Many Scottish-breeding snipe move to Ireland, while immigrant birds arrive from the north and east.

BLACK-TAILED GODWIT

Limosa limosa
Cearra-Ghob

Identification The black-tailed godwit can be distinguished by its conspicuous white wing bars and straight bill. In spring many birds show signs of their chestnut breeding plumage, especially on head and neck.

When & where It winters in small numbers on the larger estuaries, but is much more widely scattered on muddy shores during migration periods.

BAR-TAILED GODWIT

Limosa lapponica
Roid-Ghuilbneach

Identification These long-legged waders are most likely to be confused in flight with greenshank or curlew, which also have white rumps and no wing bars.

Feeding Feed by probing with their slightly upturned bill, or in shallow water with head and neck submerged.

When & where Widely distributed on muddy and sandy shores, with large numbers wintering on the east-coast estuaries.

WHIMBREL

Numenius phaeopus
Eun Bealltainn

Identification The whimbrel is much smaller than a curlew (p.80) and has a proportionately shorter bill and a conspicuously striped crown. It is most easily recognised by its call, seven or more notes all at the same pitch and evenly spaced; this is sufficiently distinctive to allow birds flying overhead in the mist to be identified.

When & where Whimbrel breed regularly in Shetland, on heather or grassy moorland, and occasionally at a few other places in the north and the Outer Hebrides. They occur in larger numbers as passage migrants, in May and July–August.

CURLEW

Numenius arquata
Guilbneach

Identification The curlew's voice and long down-curved bill make it easy to recognise.

Habitat In the breeding season curlews frequent farmland and the lower hills, usually nesting on rough grazing or damp moorland but occasionally on fields.

Voice Its song starts with several slow deep whistling notes and then accelerates into a long liquid bubbling or rippling trill. This is one of the most familiar spring sounds in Scotland's glens. The equally familiar ringing 'coor-lee' is the usual contact call, used on the breeding ground as the birds circle and glide high in the air, in wintering areas and whilst in flight on migration.

When & where Curlews appear on the breeding grounds from late February and start to leave towards the end of June. By early August flocks have gathered at the coast, where they feed over the mud flats and roost on fields nearby. Very large numbers winter on farmland in Orkney; elsewhere they are dependent upon being able to probe into estuarine mud for food.

REDSHANK

Tringa totanus
Maor-Cladalch

Identification Their long bright orange-red legs, conspicuous white wing bar and rump, and noisy calls make redshanks easy to identify. They fly fast, tilting from side to side, holding their wings upraised after landing.

Voice A variety of yelping sounds; the commonest is 'tyu yu-yu', with emphasis on the first syllable. Make yodelling and 'yip-yip' calls on breeding grounds.

When & where Breed on low moorland and damp pasture, and winter on the coast on the larger estuaries.

GREENSHANK

Tringa nebularia
Deoch-Bhiugh

Identification Distinguished from the redshank by the absence of wing bars, and by its call, a deliberate ringing trisyllabic 'tyu-tyu-tyu', usually in flight.

When & where Breeds on damp moorland with small pools and lochans, mainly in the northwest. It sometimes feeds by the shores of larger lochs, and when moving south in July–October regularly appears on estuaries.

TURNSTONE

Arenaria interpres
Trilleachan Beag

Identification The turnstone's pied flight pattern is distinctive among small waders and easily separates it from the purple sandpiper, which also frequents low rocky shores. Turnstones wintering here have predominantly grey-brown backs but these are often speckled with chestnut, especially in autumn and spring. They are gregarious and quarrelsome and give frequent staccato 'tuk-a-tuk' calls as they turn over stones and seaweed with their short bills.

When & where Turnstones are most abundant in the east and north.

Waders

RED-NECKED PHALAROPE

Phalaropus lobatus
Deargan Allt

Identification
Phalaropes swim
high in the water,
twirling round as
they grab insects on the surface, and come onto land
only at the nest. The female is more brightly coloured
than the male, which is responsible for incubating the
eggs and rearing the chicks.

When & where There is a small colony in Shetland, on
the RSPB's Fetlar Reserve, but otherwise only a very few
nesting pairs. Phalaropes summer on lochs and
marshes with small pools and winter at sea.

DOTTEREL

Charadrius morinellus
Amadan-Mòintich

Identification Males, which
incubate the eggs and rear the young, are duller in
colour than females. They are
inconspicuous and very
tame, often allowing a
close approach. When flushed,
they give a quiet trill.

When & where Dotterel breed on bare
mountain plateaux, mainly in the
Cairngorms and Grampians.

GREAT SKUA

Stercorarius skua
Fàsgadan

Identification Their heavy build, darker brown colour, conspicuous white wing patches and short tails distinguish great skuas from young gulls and arctic skuas. They harass for their fish catch, and sometimes kill, birds up to gannet size. At their moorland colonies, most of which are in the Northern Isles, the breeding birds dive-bomb intruders. 'Clubs' of immatures gather near freshwater pools. Great skuas often stand with raised wings and have a gruff 'uk-uk-uk' call.

When & where Away from the breeding colonies, they may be seen moving south off mainland coasts in August–September.

ARCTIC SKUA

Stercorarius parasiticus
Fàsgadair

Skuas, Gulls and Terns

Identification A dark-backed 'gull' with extended central tail feathers, this piratical bird flies fast and gracefully, twisting and turning, as it chases kittiwakes and terns and dives to catch the fish they drop. Near its moorland nest it pretends to have a broken wing to distract attention from egg or chick and dive-bombs intruders aggressively. This species occurs with white, dark and intermediate underparts.

Voice Its most common calls are a sharp 'ya-wor' when alarmed and a wailing 'eee-air'.

When & where Arctic skuas breed mainly in the Northern Isles; in autumn they pass down both east and west coasts on migration.

COMMON GULL

Larus canus
Faoileann

Identification This gull is similar in plumage pattern to a herring gull (p.91) but is much smaller and has no red spot on its greenish-yellow bill. It can be distinguished from the slightly smaller black-headed gull by the absence of any coloured feathering on its head. Its call is a squealing 'keeya', rather high and shrill.

Habitat In summer, common gulls are widely distributed, nesting colonially on rocky islands, on shingle banks along coasts and rivers, on moorland bogs and beside hill lochs with stony shores. In winter they desert the moorlands and move to farmland and estuaries, roosting on nearby lochs and reservoirs or sandy beaches.

Feeding When feeding, common gulls run behind the plough, make short leapfrog flights as they hunt for worms and insects on grassland, and pick prey off the surface of water.

When & where In late summer and autumn large numbers of immigrants from the continent arrive and by mid-winter there are flocks of several thousand at various places near the east coast.

BLACK-HEADED GULL

Larus ridibundus
Faoileag
Dhubh-Cheannach

Skuas, Gulls and Terns

Identification This is the least sea-frequenting of the gulls and by far the most tame and the most common inland. Its red bill and legs, and in summer its chocolate-brown head, are distinctive; in winter its head is white with smudgy dark spots above and behind the eye. In flight the white front edge to the wing makes it look paler than a common gull. Newly fledged young are a scaly red-buff above, and first-winter birds have darkish bars on their wings.

Habitat This species breeds colonially on lowland bogs, marshy loch edges and sand dunes, always near shallow calm water. It often nests among floating vegetation and the chicks have to swim almost as soon as they hatch. They can often be seen hawking over grassland for flying insects such as craneflies and ants.

Voice Black-headed gulls are gregarious and noisy all year; their usual call is a harsh scolding 'karr'.

When & where This species winter on farmland, around estuaries and near towns, readily coming to parks, gardens and scavenging at picnic areas.

LITTLE GULL

Larus minutus
Crann-Fhaoileag

Identification Adults look rather like black-headed gulls (p.87) but have jet-black (not brown) heads, no black on the wings and dark underwings. Blackish zigzags on the back make young birds resemble young kittiwakes (p.92) which are not normally found in the same habitat.

When & where Small parties of little gulls visit the east coast on migration. They are most regularly seen from Angus south, hawking for insects over freshwater lochs, less often on the coast itself or adjoining moorland.

GREAT BLACK-BACKED GULL

Larus marinus
Farspach

Identification This is the largest and darkest gull. Its legs are pale pink. They are predatory and rather vicious, killing adult and young birds, and often feed on carrion.

When & where It breeds most abundantly in the north and west but some nest on the east coast south of Aberdeen. Great black-backs prefer breeding sites not easily accessible to man and rarely nest in large colonies. In winter they occur all round the coast, with the biggest gatherings around fishing ports or near coastal rubbish tips. They are seen less inland than the herring gull (p.91).

LESSER BLACK-BACKED GULL

Larus fuscus
Farspach Bheag

Identification Lesser black-backs have yellow legs.

Habitat They nest on rocky headlands, rough ground on small islands and moorland rather than clifftop positions.

When & where Most breed along rocky coasts in the Northern and Western Isles and on the northwest mainland, though there are colonies on the Isle of May and inland at Flanders Moss. Most move away from Scotland in winter; when returning from the south in February and March lesser black-backs often visit farmland.

HERRING GULL

Larus argentatus
Glas-Fhaoileag

Identification The palest, commonest and most widespread of the big gulls. Breeding pairs are noisy and often aggressive and may dive-bomb intruders. This behaviour, together with their loud and strident early-morning 'kyow' calls, makes them decidedly unpopular with human neighbours!

Habitat This species breeds on all rocky coasts, occasionally inland and on rooftops in seaside towns.

Feeding Herring gulls scavenge almost any edible refuse and regularly follow boats at sea.

When & where In winter the largest concentrations are around east-coast fishing ports and near garbage tips in the central lowlands.

KITTIWAKE

Rissa tridactyla
Seagair

Identification Much the smallest of the cliff-nesting gulls, the kittiwake has short black legs, a yellow-green bill, and a red mouth which is easily seen when it calls 'kitti-wak'.

Habitat The kittiwake is a maritime species, coming to land only at its colonies, where its mud and seaweed nest is attached to a tiny ledge on the cliff face.

Feeding Kittiwakes plunge-dive and often gather in flocks over shoals of small fish well offshore.

When & where They are most abundant in the Northern Isles and on suitable north and east-coast cliffs, scarcer in the west.

ARCTIC TERN

Sterna paradisaea
Steàrnal

Skuas, Gulls and Terns

Identification This tern's bill is blood-red right to the tip, whereas a common tern's has a black tip. When seen overhead the points of its wings look semi-transparent. It is often difficult to be certain of sight identification, but location is a useful guide to the most likely species.

Breeding They nest colonially on rocky islands and coastal moorland and are aggressive, often dive-bombing intruders.

When & where Like all the terns, these are summer visitors. They are by far the most abundant terns in the Northern Isles, more thinly scattered in the west and relatively scarce on the east coast.

COMMON TERN

Sterna hirundo
Steàrnag

Identification A black tip to its orange-red bill, longer legs and a shorter tail are the most obvious differences between this species and the arctic tern. Only a small patch on the wing looks semi-transparent from below, not the whole tip.

Habitat Common terns nest on sand and shingle beaches, and occasionally inland on river shingles.

When & where They breed in almost every county and are the commonest terns in the southwest and in the east from Aberdeen southwards. Their colonies seldom exceed 100 nests, whereas more than 1,000 pairs of arctic terns sometimes occupy a site.

SANDWICH TERN

Sterna sandvicensis
Steàrnag Mhòr

Identification This is the largest and most gull-like of the terns and is easily distinguished by its untidy black crest and mainly black bill.

When & where It is much more scarce than common and Arctic terns and breeds at only a few sites. The most regularly occupied colonies are in Orkney, East Ross, Grampian – at Sands of Forvie and the Loch of Strathbeg – and on the island of Inchmickery in the Forth. Sandwich terns, like the other species, gather on beaches in July and August before migrating.

Scottish Birds

LITTLE TERN

Sterna albifrons
Steàirdean

Identification Its small size, mainly yellow bill and white forehead readily identify the little tern, which is less common than the other species.

Breeding Its colonies are usually on sandy beaches, and seldom comprise more than a few pairs. Because they choose beaches attractive to man, and frequently lay close to the high tide mark, little terns often fail to breed successfully unless given special care and protection on wardened reserves.

When & where Widely but very thinly scattered colonies.

GUILLEMOT

Uria aalge
Eun Dubh an Sgadain

Auks

Identification Guillemots have slender pointed bills; some have a white 'bridle' round the eye. At their colonies they keep up a noisy trumpeting 'aargh' and form rafts on the sea nearby.

Feeding Guillemots surface-dive for fish and swim fairly high in the water.

Breeding They pack close together on their breeding ledges and face towards the cliff to prevent eggs or chicks from falling into the sea. However, fledglings still leave before they are able to fly.

When & where They are most abundant in the north and east. They winter at sea and, sadly, some oiled birds are found on beaches most winters.

Scottish Birds

BLACK GUILLEMOT

Cepphus grylle
Gearradh-Breac

Identification

This is by far the smallest and most scarce of the auks, and is readily identified by its white wing patch, conspicuous at rest and in flight. In winter they look very different, with mottled grey-white back and sides.

Breeding They nest solitarily, in rock crevices and on boulder beaches, but gather to display early in the breeding season.

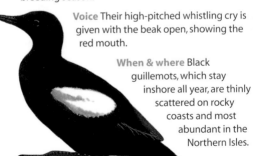

Voice Their high-pitched whistling cry is given with the beak open, showing the red mouth.

When & where Black guillemots, which stay inshore all year, are thinly scattered on rocky coasts and most abundant in the Northern Isles.

RAZORBILL

Alca torda
Coltraiche

Identification Razorbills have a conspicuous white vertical line near the tip of the deep bill and are blacker than guillemots. When taking off from the water they patter along the surface.

Breeding They usually nest singly, in cliff crevices or among boulders, and can often be seen offshore in family parties in late summer. They winter at sea.

When & where
They are much less numerous than guillemots and, like them, are most abundant in the north, though breeding on suitable cliffs right round the coast.

PUFFIN

Fratercula arctica
Fachach

Identification Their brightly coloured bills and white cheeks make puffins unmistakable in summer, as they stand around near their nesting burrows, which are usually on steep grassy slopes. Their bills lose much of their colour in winter, which is spent far out at sea. Puffins gather in rafts on the sea near their colonies, looking even dumpier on the water than ashore, and fly to and fro with rapidly whirring wings.

When & where
They are most
common on islands
in the north and
scarce south of
Caithness, except
on the Isle of May.

WOODPIGEON

Columba palumbus
Calman Coille

Identification This is the only pigeon with a white patch on the side of the neck (absent in young birds) and a broad white crescent, conspicuous in flight, from front to back on the wing. They are noisy birds, 'exploding' when disturbed with a loud clatter of wings; their display flight is also noisy, involving loud wing claps.

Voice The woodpigeon's song is a lazy-sounding 'coo cooo coo coo-coo', with the emphasis on the second note; the stock dove (p.103) stresses the first syllable.

When & where Woodpigeons are widely distributed and abundant; they sometimes damage crops and can be a serious pest, especially where woods close to farmland attract large roosting flocks in winter.

COLLARED DOVE   *Streptopelia decaocto*
Calman Coilearach

Identification This elegant, small pigeon is very closely associated with man and is a frequent and tame garden visitor. The white tips to its outer tail feathers are very obvious in flight or when the bird fans its tail while sunning itself. It flies fast, with looping falls, glides and dives, bobs its head when walking, and frequently perches on TV aerials, lamp posts and roofs.

Breeding They have an unusually long breeding season, allowing them to raise several broods a year, which has doubtless helped their rapid spread.

Voice The collared dove's distinctive 'ku-kroo-ku' song, with the accent on the middle syllable, is occasionally mistaken for the call of the cuckoo; it has a harsh, nasal 'kwarr' flight call.

When & where Collared doves first appeared in Scotland as recently as 1957. They have been seen as far north as Shetland, but are commonest around towns, villages and farms in the central lowlands and up the east coast to Dornoch.

STOCK DOVE

Columba oenas
Calman Gorm

Pigeons

Identification Stock doves lack the woodpigeon's white marks and can be confused with feral pigeons (p.104), which do not always have white rumps. They often feed with woodpigeons in winter but are less numerous and likely to be seen only in small groups.

When & where Stock doves are found on low ground from the central lowlands southwards and up the east coast to the Moray Firth. They feed among trees and shrubs as well as in the fields, but do not occur in town parks or gardens.

ROCK DOVE

Columba livia
Calman Creige

Identification Many feral pigeons look like rock doves, from which they are descended. They are smaller than woodpigeons (p.101) and have no white on the neck or wing; in flight their white rump and double black wing bar are conspicuous.

When & where Pure rock doves are now found only on the north and west coasts, breeding in caves on both mainland and islands. Rock doves seldom perch in trees, preferring to settle on the ground, rocks or cliff ledges.

CUCKOO

Cuculus canorus
Cuthag

Identification Although the cuckoo's two-syllable call is familiar, the bird itself is often unrecognised. It flies low and fast, usually escorted by several small birds, and looks rather like a sparrowhawk (p.49) but has more pointed wings and white tips to its long tail. It perches almost horizontally on posts or branches, more upright on wires.

Breeding Cuckoos breed on moorland and heath and near open woodland. In Scotland they usually lay their eggs in the nests of meadow pipits (p.119).

When & where They are widespread, but most abundant and obvious in the northwest and the islands.

NIGHTJAR

Caprimulgus europaeus
Sgraicheag Oidhche

Identification This nocturnal, moth-eating summer visitor is more likely to be heard than seen. It sings at night, with a loud fast churring which rises and falls and is often sustained. During the day it sits motionless on a branch or the ground, where its colouring provides perfect camouflage.

Habitat They show a preference for rather open woodland, either deciduous or coniferous, with rhododendron or bracken and other ground cover in glades and rides.

When & where Nightjars occur regularly only in Dumfries and Galloway and on Arran. This species was more widespread in the past; its decline is believed to be partly due to climatic change, with late cold springs adversely affecting the availability of moths.

BARN OWL

Tyto alba
Cailleach-Oidhche

Owls

Identification Barn owls are most likely to be seen in the beam of headlights as they hunt by night along a roadside, or roosting by day in a building. They are largely nocturnal but sometimes hunt in late afternoon. This is the palest of the owls, with a large round head, a heart-shaped white face and dark eyes. Its long, white-feathered and rather 'knock kneed' legs dangle when it flies and are clearly visible as it stands upright on a perch. Its hunting flight is low, wavering and absolutely silent. At the nest it hisses and 'snores'; its main call is an eerie shriek.

Breeding They nest in tree-holes or on a ledge inside a building; on some farms special shelves have been put up in modern barns to provide suitable sites.

When & where Barn owls are most abundant in the Borders and southwest. They are rather scarce throughout the central lowlands and up the east coast and absent from most of the highlands and islands.

TAWNY OWL

Strix aluco
Comhachag Dhonn

Identification Tawny owls are tubby, with large round heads, and are a rather uniform tawny or greyish brown, with paler facial 'discs'. Tawny owls are more often heard than seen, as they normally hunt between dusk and dawn, dropping silently onto their small rodent prey. Pellets of indigestible skin and bones on the ground beneath a tree show that owls have been roosting there.

Habitat Primarily a woodland resident, it also frequents parks and suburban gardens. Where nesting holes are scarce, they occupy boxes.

Voice This is the commonest owl and is responsible for the familiar 'too-whit' and 'hoo-hoo-hoo' calls. It defends its territory throughout the year and regularly calls in October and November.

When & where Widespread on the mainland, they are abundant in mature deciduous or mixed woods.

LONG-EARED OWL

Asio otus
Comhachag Adhairceach

Identification A mainly nocturnal hunter, this owl is more likely to be found at its roost than seen in flight. When disturbed at the roost it draws itself up like a cat, making its body appear taller and slimmer, while its glowing orange eyes and conspicuous ear-tufts distinguish it from the tawny owl (p.108).

Habitat Long-eared owls typically occupy small patches of conifer woodland surrounded by open hunting habitat, often moorland. They usually lay in the old nests of other species, such as crows.

Voice The long-eared owl's cry is a soft, moaning 'oo-oo-oo'; it is a much more silent bird, even during the breeding season, than the tawny owl. Young birds, when nearly or newly fledged, give loud squeaky calls which sound rather like a creaking gate; these are often the first sign of this species' presence.

When & where Widespread on the mainland.

SHORT-EARED OWL

Asio flammeus
Comhachag Chluasach

Identification Short-eared owls hunt by day, with wavering, moth-like wing beats and short glides with wings upraised. When circling in display flight they frequently clap their wings together below their bodies and 'sing' with a deep 'boo-booboo'. They perch in a crouching position, on the ground, a rock or a fence post and at close range look fierce, due to their glaring yellow eyes.

Breeding Short-eared owls breed on moorland and recently afforested ground, where they feed on small rodents; breeding success is closely linked to the size of the vole population. Like many predatory birds, they start to incubate when the first egg is laid, so the chicks hatch out at intervals. In a good year for short-tailed voles all the young owls are likely to be reared; in a bad year only the oldest one or two chicks will survive.

When & where This is the only owl to breed regularly in Orkney and the Outer Hebrides. In winter many short-eared owls move south; those that remain hunt over lower ground and coastal areas.

KINGFISHER

Alcedo atthis
Biorra-Crùidein

Identification Usually seen as a vivid flash of blue with whirring wings as it heads away low and fast over the water, the kingfisher is unlikely to be confused with any other species. When seen side-on in flight or perched, its chestnut underparts are very striking.

Feeding It dives head-first when fishing, plunging in either from a perch, where it may bob nervously as it watches the water for prey, or after hovering briefly in flight. It often returns to the same perch between dives, and usually batters larger fish against the perch before swallowing them.

Breeding Kingfishers suffer badly in severe weather, when rivers freeze, and a few hard winters in succession greatly reduce the population. The fact that two broods are raised in good years helps the species to recover quickly after such a setback.

When & where A bird of slow-flowing lowland rivers, the kingfisher is thinly scattered as far north as the Moray Firth. Its distribution is largely determined by the availability of small fish and it sometimes takes advantage of fish farms.

GREAT SPOTTED WOODPECKER

Dendrocopos major
Snagan Daraich

Identification The great spotted woodpecker's loud territorial drumming, produced by rapidly pecking at a branch, is often the first sign of its presence and immediately identifies it. Its boldly pied plumage is also distinctive and in flight, which is markedly undulating, its white shoulder patches and red undertail are conspicuous. Females lack the male's red nape patch, and young birds have red crowns. This species uses its tail as a prop as it chips away at dead wood to get at insects; scatterings of fresh wood chips show where a woodpecker has recently been at work. A sharp 'chik' call is given often, both while clinging to a tree and in flight.

Habitat They frequent both deciduous and coniferous woodland, but can only breed where there are dead trees in which they can excavate their nest holes. In hard weather they sometimes visit bird tables for fat.

When & where Widespread on the mainland and also occur on a few of the larger, well-wooded islands.

GREEN WOODPECKER

Picus viridis
Lasair Choille

Identification This is the only medium-sized bird which is predominantly green on back and wings. Its yellow-green rump is conspicuous in flight and its crimson crown is also obvious. Green woodpeckers seldom drum; instead they make their presence known by their loud, carrying 'yaffle' call, which sounds like maniacal laughter and gradually accelerates as it descends the scale. Like the great spotted woodpecker, this species flies in deep undulating loops, closing its wings for several seconds between upward bounds. On the ground it holds itself rather upright and moves in hops.

Feeding They feed mainly on the ground and are especially fond of ants.

When & where A bird of low-ground deciduous woodlands, the green woodpecker requires a mix of mature trees and open ground. It started to breed in Scotland only in the 1940s and has since gradually extended its range; it now occurs as far north as Rossshire and has nested on Islay.

PIED WAGTAIL

Motacilla alba
Breac an t-Sìl

Identification This is the only small ground bird with predominantly black and white plumage and a long tail. Females and young birds are greyer on the back and in winter all pied wagtails look markedly whiter in the face, as they lose much of the black bib, with only a crescent-shaped 'scarf' remaining.

Habitat Pied wagtails are widespread in towns and gardens, around farm steadings, and in open country, wherever there is a good supply of insect food and of nesting sites: holes in walls or buildings, crevices in steep banks, or thick ivy.

Feeding They run about picking up insects and making short fly-catching flights.

Voice A brisk 'chizzik' is the commonest call; it is often given in flight and is quite distinctive.

When & where Between August and March pied wagtails join up in flocks and roost communally, in reedbeds or among willows. In severe weather they often move into little-used buildings for added protection.

YELLOW WAGTAIL

Motacilla flava
Breacan Buidhe

Larks, Pipits and Wagtails

Identification Yellow wagtails have green, rather than grey, backs and no throat markings, unlike the longer-tailed grey wagtail (p.116).

Habitat The yellow wagtail's habitat is also quite different; it favours damp meadows, marshy ground, hayfields and occasionally cornfields, whereas the grey wagtail is very much a bird of fast and rocky streams and rivers.

When & where Yellow wagtails nest regularly only in a few places in low-lying parts of Ayrshire, Lanarkshire and the Borders.

GREY WAGTAIL

Motacilla cinerea
Breacan-Baintighearna

Identification The grey wagtail is more contrasted in colour than the yellow, and is longer and slimmer-looking due to its very long tail, which is constantly whipped up and down. It is restless and active, running over stones to pick up insects and flycatching, from a perch or the ground, with fluttering leaps and hovers. Its direct flight is bounding and it often gives its shrill 'siz-eet' as it flies.

Habitat The grey wagtail is typical of hill streams and fast rocky lowland rivers, a quite different habitat from that of the yellow wagtail (p.115). They are most abundant where there are rapids or broken water.

When & where In summer, grey wagtails are widely distributed on the mainland and most of the large inner islands but scarce in the Outer Hebrides and Northern Isles. In winter most move no further than England and a few remain in the central lowlands.

SKYLARK

Alauda arvensis
Uiseag

Identification The skylark's sustained and musical song, delivered in high-level, hovering song flight, is its most distinctive feature. Crouching on the ground, or on a fence post, it looks decidedly dumpy and its short crest is not always obvious. A liquid 'chirrup' is often given as it flies, when wing beats alternate with closed-wing glides, resulting in an undulating flight pattern.

When & where Skylarks are the most widely distributed British species, breeding on farmland, rough grassland, heather moors and hilltops, from sea level to over 1000m. In winter they desert high ground and move nearer the coast.

TREE PIPIT

Anthus trivialis
Riabhag Choille

Identification Differences in voice and choice of habitat help to distinguis this species from the similar meadow pipit (p.119). Its plumage pattern is like the meadow pipit's but it is brown rather than olive above and yellower on the breast, which is less heavily streaked. Young birds are more buff-brown on the back than adults.

Habitat Tree pipits are found in open woodland or among scattered birch, pine or mixed deciduous trees. Clearings within the wood, and tall trees from which to launch into song flight, are essential.

Voice The tree pipit has a loud and musical song, more varied than the meadow pipit's, incorporating long trills and ending with a slow 'chew-chew-chew' or 'seea-seea'. It sings from the top of a tree or, more often, in song flight, which involves flying upwards from a perch, circling and parachuting back down.

When & where Although widely distributed over much of the mainland, this species is scarce in the arable farming areas of the central lowlands and east coast.

MEADOW PIPIT

Anthus pratensis
Snàthtaq

Identification The meadow pipit is streaky-brown with white outer tail feathers, similar to the skylark (p.117), but it is smaller and slimmer and has a very different voice. Its call is a thin 'tseep' and its song, given during its flutter-up and parachute-down song flight, is a series of thin piping phrases, ending in a trill.

When & where Mainly a bird of open ground. Many move south in winter.

ROCK PIPIT

Anthus petrosus
Ghabhagan

Identification Darker than a meadow pipit, with grey, not white, outer tail feathers, similar thin 'seep' calls and a less tuneful song consisting largely of 'see' sounds. Rock pipits do not flock, even where there are many present.

When & where A bird of tidal rocky shores, especially in the north and west, present throughout the year.

SWALLOW

Hirundo rustica
Gobhlan-Gaoithe

Identification The swallow's long tail streamers distinguish it at once, whether in the air or at rest. Young birds are duller in colour and have shorter streamers. Swallows look short-legged and shuffle awkwardly when on the ground, but are swift and agile flyers; at the nest they often hover with tail fanned. They sing a lot, both in flight and when perched, with a pleasant twittering warble.

Habitat Most swallows frequent farmland, where they nest on rafters in buildings and often return year after year to the same site.

Feeding They feed mainly by catching insects in flight, usually over water or near animals, but sometimes pick them off roofs or plants.

When & where They start to arrive in early April. Swallows are much less sociable than house martins during the breeding season but before they set off for Africa in late summer large flocks, of several thousand birds, gather in various parts of the country.

SWIFT

Apus apus
Gobhlan Mòr

Swallows, Martins and Swifts

Identification The swift's long, narrow, rather sickle-shaped wings, shrill screams and fast, dashing flight make it an easy bird to identify. These are truly birds of the air; they feed and mate on the wing and land only at the nest site. In flight they hold their wings stiffly and beat them very rapidly. The harsh screaming calls are given mainly in flight, but also sometimes by birds on the nest.

Breeding They nest in buildings, usually under the eaves, but will also occupy specially designed nesting boxes.

When & where Swifts are entirely dependent upon flying insects for food and in consequence are among the latest of the summer visitors to arrive, most reaching Scotland about mid-May. From then until early August 'screaming parties' are a familiar sight, but as soon as the weather turns cool the swifts depart. Swifts are most common in lowland areas, rather local in the highlands and absent from the islands. In warm spells they are sometimes seen high in the mountains.

HOUSE MARTIN

Delichon urbica
Gobhlan-Taighe

Identification Its broad white rump and relatively short, shallowly forked tail distinguish the house martin from the swallow. When on the ground, or clinging to a nest, the white feathering on the legs is conspicuous. They waddle, often with wings and tail raised. They are sociable, feeding and gathering mud for nest-building in small parties, often building nests close together under the eaves of the same house.

Habitat Most house martins are associated with towns and villages, although small colonies also occur on more isolated buildings and some nest on sea and inland cliffs.

Voice House martins have a soft and sweet twittering song; their call is a clear but rather quiet 'chirripp'.

When & where Like most species which feed upon flying insects, house martins do not reach Scotland until late April or even early May. They are scarce over much of the highlands and islands. In late summer, flocks gather in preparation for migration and move south in waves, especially near the east coast, on their way to Africa for the winter.

SAND MARTIN

Riparia riparia
Gobhlan Gainmhich

Swallows, Martins and Swifts

Identification This is the only small brown and white bird which regularly feeds by hawking over water for flying insects. Sand martins are very gregarious and are usually seen in groups, when they keep up a rather harsh conversational twittering. Their flight is fast and agile, with frequent flutterings and changes of direction.

Breeding They nest colonially in exposed vertical faces of sand or fine gravel which are either over water or relatively high; when a bank becomes vegetated, or slips from the vertical, the site is abandoned.

When & where Sand martins arrive in late March or early April. Before leaving in late summer they gather in large flocks.

DUNNOCK

Prunella modularis
Gealbhonn nam Preas

Identification This is the least conspicuous of the regular garden residents. Mouse-like in colouring, it also behaves rather like a mouse, creeping along in an unobtrusive way. It does draw attention to itself, however, by frequent wing flicking, by its insistent, penetrating 'tseep' calls and by its song, a thin warble, much shorter and less forceful than a wren's but quite musical. Like the wren it sings almost all year round and usually makes only short flights in the open.

Habitat Dunnocks are solitary. They prefer a bushy habitat and frequent hedges and woods with a good layer of undergrowth, as well as gardens.

Feeding They eat both insects and seeds, often foraging among fallen leaves, and feed more often under bird tables than on them.

When & where Apart from the Northern Isles and Outer Hebrides, where they are scarce or absent, they are widespread.

DIPPER

Cinclus cinclus
Gobha-Dubh

Identification Its dumpy shape, brilliant white shirt front, and habit of crouching and bobbing its whole body up and down, make the dipper easy to identify. Dippers fly low and fast above the water, are buoyant swimmers and sometimes dive under ice.

Habitat It is a bird of fast-flowing streams and rivers, and also frequents stony loch shores in winter. Dippers are highly territorial and widely spaced out along suitable stretches of river.

Feeding It feeds mainly on the stream bed, either walking in from a stone or plunging into the water, always facing into the current.

Breeding They tend to use traditional nest sites, often under bridges or on a rock; the same domed nest may be used in successive years or a new one built nearby.

Voice The call is a sharp 'zit' and dippers sing their high, rather grating, warble all year.

When & where This species is widespread on the mainland and most of the inner islands.

WREN

Troglodytes troglodyte
Dreathan-Don

Identification With its cocked tail and loud voice, the diminutive wren is easy to recognise. Always on the move, it hops along the ground or climbs up stems, searching every nook and cranny for small insects and frequently flicking its tail. Much of its time is spent in the shelter of bushes and low vegetation. Wrens usually make short flights before diving into cover.

Breeding The male builds several domed nests but only the one chosen by the hen is lined. In severe winters numbers decline as they are not attracted by bird-table foods and do not move away to a warmer wintering area. To keep warm wrens sometimes roost communally in nest boxes or similar places.

Voice Wrens sing almost throughout the year – a forceful and carrying, rather rattling, warble incorporating vigorous trills. They 'churrr' loudly when alarmed, and maintain contact with hard clicking 'tik-tik-tik' calls.

When & where Wrens are found in gardens, woods and hedges, on rocky seashores and high in the hills. They seldom travel far, and as a result local 'races' have developed on some of the islands.

Wren / Thrushes and Chats

ROBIN

Erithacus rubecula
Am Brù-Dhearg

Identification Although by no means the most common British bird, the 'robin redbreast' is probably the best known. Young birds lack the red breast and are a scaly looking golden brown; by winter they are indistinguishable from adults. Robins are unusual in that during autumn and early winter both sexes defend territories. They are very pugnacious, with much fluffing out of breasts to display the red patch to best advantage. In mid-winter the birds pair and share a territory. Garden robins can be quite tame, sitting close by when digging is in progress, hoping for some titbit to be turned up.

Voice Both sexes sing, except for a short period in midsummer when song stops altogether. The song, a short and melodious warble, has a plaintive quality which is distinctive and easy to recognise. The usual calls are a sharp, scolding 'tik, tik', often used when birds are alarmed or going to roost and a thin, high 'tswee'.

When & where Widespread except in Northern and Western Isles.

REDSTART

Phoenicurus phoenicurus
Eàrr Dhearg

Identification The redstart is easily recognised by its rusty red rump and tail, which it constantly quivers up and down while on the ground and flirts fanwise in flight. Females are greyish brown above and pale orange-buff below, and juveniles are similar but speckled like young robins (p.127).

Habitat This species is found in relatively open woodlands, especially those including oaks and pines.

Feeding Redstarts are hole-nesters, but readily make use of nest boxes. They feed both on the ground and in the trees, occasionally hovering or flycatching in flight.

Voice Redstart song is a rather robin-like short warble, but weaker and ending in a feeble twitter. The commonest calls are 'hwee-tuk-tuk' and a loud 'hweet', not unlike that of a willow warbler.

When & where Redstarts are most abundant in the central, west and northwest highlands, mainly in wooded glens. It breeds only locally in other parts of the mainland and not at all on most of the islands.

WHINCHAT

Saxicola rubetra
Gocan Conaisg

Identification Both male and female have conspicuous stripes above the eye and at the side of the throat, white in males and buffish in females. White patches on the inner wing and the sides of the tail show in flight. Their short warble starts harshly but becomes sweeter.

When & where They breed in areas of rough ground with scattered bushes and in young plantations. They are more widely distributed than stonechats, especially inland in eastern Scotland.

STONECHAT

Saxicola torquata
Clacharan

Identification The male's black cap and white half-collar are distinctive. In flight both sexes can be distinguished from whinchats by the absence of white on their tails. The stonechat's song is a rapidly repeated double note, rising and falling.

When & where This species is most abundant in the west, where it is resident. It is usually found on rough ground with gorse.

WHEATEAR

Oenanthe oenanthe
Brù-Gheal

Identification The wheatear has a conspicuous white rump, which is obvious as it flies or hops around. Its song is an almost lark-like warble, but much shorter and at times rather creaky and wheezy.

When & where The wheatear occurs on all the islands and suitable areas of the mainland. A hole-nester, it frequents open ground with short vegetation and scattered stones or rocks. It is one of the earliest summer visitors to arrive.

RING OUZEL

Turdus torquatus
Dubh-Chreige

Identification The white crescent is clearer and brighter on males than females, and absent on young birds. Their loud scolding calls are similar to a blackbird's (p.131).

When & where Found in most moorland and mountainous areas of the mainland and some of the inner islands, they breed on open ground with scattered trees, often near crags or streams.

BLACKBIRD

Turdus merula
Lon-Dubh

Identification The sombrely coloured blackbird is one of the most familiar of garden residents. Young birds and females are brown rather than black and lack the adult male's bright orange bill. This species not infrequently produces pied partial albinos, which range from black with white spots or collar to almost completely white.

Feeding They feed largely on worms, but also take insects, berries and household scraps; individuals sometimes become very tame when regularly fed.

Breeding Their nests, found in bushes or hedges, are made of twigs, grass, moss and mud. Two or three broods may be raised.

Voice Blackbirds are among the first contributors to the dawn chorus; their song is rich and melodious, with each phrase lasting for several seconds and delivered at a leisurely pace. Alarm calls include a screeching chatter and a loud 'chink-chink'.

When & where Occur almost everywhere there are trees, except in highland birch woods and conifer woodlands.

FIELDFARE

Turdus pilaris
Liath-Truisg

Identification The grey head and rump immediately distinguish this winter visitor. They are noisy, with a frequent harsh 'chak-chak-chak' call.

Habitat They feed in the open on fields, but also take berries from hedgerows. In hard weather fieldfares come into gardens, and readily take fallen apples.

When & where Widespread in winter.

REDWING

Turdus iliacus
Sgiath-Dheargan

Identification Redwings can be distinguished by their cream eye stripe and the chestnut on flank and underwing. Their call is a high thin 'see-ip'.

When & where This species is predominantly a winter visitor, though a few pairs breed in this country. Redwings roam over farmland and hedgerows, often in company with fieldfares.

SONG THRUSH

Turdus philomelos
Smeòrach

Identification This spotty-breasted species is shyer than the blackbird (p.131), spending much time in or close to cover, and it is a less frequent visitor to bird tables. When feeding it typically runs or hops for a short distance, then stands with head cocked.

Feeding In addition to worms, insects and berries this species eats snails, which it breaks by hammering the shell on a stone.

Breeding Its nest is similar to a blackbird's but without the mud.

Voice Its varied and musical song, often delivered from a TV aerial or tree top, is made up of short phrases, each repeated two or three times, with pauses between. Its usual call is a thin 'sipp' or 'tick'.

When & where The song thrush is widely distributed wherever there are shrubs or trees, but less abundantly than the blackbird. In autumn, many move to warmer climes. Hard winters threaten those which remain with starvation, as frozen ground and snow cover prevent them from obtaining enough food.

MISTLE THRUSH

Turdus viscivorus
Smeòrach Mhòr

Identification The mistle thrush is more grey/brown than the smaller song thrush (p.133) and has more distinct large spots on its whitish breast; in flight the white 'corners' to its tail show well. The absence of grey on the head and rump distinguishes it from the fieldfare (p.132), which is similar in size. It looks bold and alert and stands rather upright. It sings from tree tops, loudly repeating a number of short phrases; the fact that it sings in all weathers has earned it the name 'storm cock'.

Habitat It is at home in any habitat where there are both trees and good expanses of open ground. It breeds in parks and large gardens as well as on farmland with small woods or scattered trees and on low moorland, but numbers are small and pairs usually well spaced out.

When & where After the breeding season mistle thrushes join up in small flocks; some leave the more exposed upland breeding areas in winter but most remain near their home ground.

GRASSHOPPER WARBLER

Locustella naevia
Ceileiriche Fionnan Feòir

Identification Both this species and the sedge warbler (p.136) have creamy eyestripes, but the grasshopper warbler's is faint whereas the sedge warbler's is very conspicuous. They also differ in tail shape, the grasshopper warbler's being rounded and the sedge warbler's graduated to a point.

Voice This streaky-brown warbler is much more likely to be heard than seen. Its song is a high-pitched mechanical churring rather like the winding of a fishing reel, frequently sustained for over one minute. Grasshopper warblers sing both day and night, and spend most of their time skulking amongst low vegetation in young conifer plantations, reedbeds, or rough marshy ground.

When & where Although they may be present until well into August, grasshopper warblers effectively 'vanish' when they stop singing in late July. They breed regularly from the central lowlands southwards, are scarcer elsewhere on the mainland, and nest at least occasionally on some of the inner islands.

SEDGE WARBLER

Acrocephalus schoenobaenus
Glas-Eun

Identification This skulking, streaky-brown warbler shows an unstreaked tawny rump in flight, and its creamy eye stripe is conspicuous when it perches in the open. It spends much of its time in thick cover, where it climbs agilely as it hunts for insects.

Habitat Sedge warblers usually breed near water, in reedbeds, beside overgrown ditches or among willow thickets, and occasionally in young conifer plantations.

Voice It sings loudly and vigorously, with a mixture of musical and harsh, chattering phrases, including various harsh 'churrs' and 'chuks', and often mimics other species. It sings day and night, and is consequently sometimes mistaken for a nightingale (which only rarely occurs in Scotland). It also sings in a short, vertical display flight.

When & where Widely distributed in the southern half of the country but rather local further north and absent from Shetland and the Outer Hebrides.

WHITETHROAT

Sylvia communis
Cealan Coille

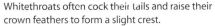

Identification The whitethroat's brilliant white throat, rusty wings and white outer tail feathers immediately distinguish it from the other common warblers. The male's head and nape are grey and the female's brown. Whitethroats often cock their tails and raise their crown feathers to form a slight crest.

Habitat Typically includes patches of tangled vegetation, brambles and small bushes set in fairly open ground or along woodland edges. This species can be difficult to spot as it spends much of its time in cover, searching for insects among foliage. Its voice is often the first indication of its presence.

Voice They have rather harsh voices, with a variety of scolding 'charr' and repeated 'chak' notes, and the male's song, often given in a brief dancing flight, is a vigorous, scratchy chatter.

When & where Whitethroats are most abundant in the central lowlands and southern counties. They are among the later-arriving summer visitors.

GARDEN WARBLER

Sylvia borin
Ceileiriche Gàrraidh

Identification This is a shy and solitary bird, which skulks in cover and seldom flies in the open long enough to give a good view. Garden warblers are rather plump and have no distinctive markings, but can be distinguished by a process of elimination! Unlike other warblers, they seldom flick their wings and tails.

Habitat They require dense shrubby cover in their breeding area and frequent patches of rhododendron or bramble growing as an underlayer in fairly open woodland, overgrown hedges and occasionally young conifer plantations at the thicket stage.

Voice Its song, delivered from cover, is important for identification – a sweet, mellow and flowing, rather subdued but well-sustained warble, sometimes lasting for several minutes with only brief pauses. They have 'chek', 'chak' and 'churr' calls.

When & where They are common from the central lowlands southwards, especially towards the west, and occur occasionally up the east coast to the Moray Firth. It is one of the later-arriving of the insect-eating summer visitors.

BLACKCAP

Sylvia atricapilla
Ceann-Dubh

Identification Their distinctive coloured caps and slim grey-brown bodies make blackcaps easy to identify, though males might be confused with the black-chinned and much plumper marsh and willow tits (p.147).

Habitat Mature deciduous or mixed woodland with good shrub cover, a combination typically found in parks and estate policy grounds.

Voice Blackcaps spend a lot of their time in cover or high in the tree canopy, and their song is often the first indication of their presence. They have a loud and carrying warble, clearer and more varied than a garden warbler's, but usually less sustained and in shorter phrases. The usual contact call is a sharp 'chak' or 'churr'.

When & where They are most abundant in the central and southern lowlands, regular up the west coast and occur occasionally in the Great Glen and near the east coast. Although essentially summer visitors, a few blackcaps are present most winters.

CHIFFCHAFF

Phylloscopus collybita
Caifear

Identification
This small warbler is
most surely identified by
its song, which is
composed of two distinct
notes, 'chiff' and a lower 'chaff',
repeated monotonously but
in variable sequence. In
appearance it is very like
the more common willow warbler (p.141), but slightly
more brownish than greenish above and buff rather
than yellowish below.

Habitat Chiffchaffs favour woods and parkland where
tall trees, either deciduous or coniferous, are
intermixed with shrubby growth; they spend much of
their time high in the canopy.

Feeding They are restless and active feeders, hopping
and flitting along the branches with constantly flicking
wings and tails, and occasionally hovering or making
short flycatching flights.

When & where This species is scarcer and more local
than the willow warbler, breeding mainly in the
southern half of the country. Chiffchaffs are the first of
the warblers to arrive in spring.

WILLOW WARBLER

Gavia stellata
Gael: Learga Mhòr

Warblers

Identification This little 'leaf' warbler is responsible for much of the bird song to be heard in Scottish woodlands in summer. Its song immediately confirms its identity and is easy to recognise: a rather plaintive sequence of silvery liquid notes descending the scale, faint at the start and becoming more deliberate. Its contact call is a soft 'hooeet'. They feed more often at low level than do chiffchaffs (p.140). During the autumn migration, young birds tend to look much brighter in colour, with decidedly yellow underparts.

Habitat Willow warblers are found in all types of woodland, except well-grown conifer plantations, but are especially typical of birch woods and are often present even where there are only a few scattered birches on a hillside.

When & where They breed throughout the mainland and on many of the islands, but not regularly in the Northern Isles. During the autumn migration, from July onwards, waves of willow warblers drift southwards, often turning up in gardens and town parks where they do not breed.

WOOD WARBLER *Phylloscopus sibilatrix*
Conan Coille

Identification The wood warbler is the largest and most green and yellow of the three small 'leaf' warblers, and has a distinct eye stripe. They spend most of their time high up in the trees, but, like other 'leaf' warblers, build a domed nest on the ground.

Habitat This species prefers mature deciduous woods with little ground cover, and is especially associated with oak woods.

Voice It is unusual in having two quite different songs, both of which are distinctive enough to identify it. One is a sustained shivering trill, which starts slowly and gradually accelerates, and the other a deliberate piping 'pyu-pyu-pyu' repeated seven or more times. Both are loud and penetrating, carrying quite a distance and rapidly disclosing their presence.

When & where It is most abundant in west central Scotland. Elsewhere on the mainland it occurs only very locally and it is absent from most of the islands.

GOLDCREST

Regulus regulus
Crionag Bhuidhe

Identification The goldcrest is the smallest British bird. It is dumpy-looking with no obvious neck, a tiny bill, very large eyes for its size and a distinct wing bar. Young birds lack the golden 'crest' and might be mistaken for a willow warbler (p.141), which is longer and slimmer, and has no wing bar.

Habitat Goldcrests are found in most types of woodland, but are most abundant in conifer woods. In winter, they often join flocks of tits and treecreepers, roaming through the trees in search of minute insects. Hard winters can result in many deaths, as these tiny birds are unable to find sufficient food during daylight to sustain them.

Voice Goldcrests spend much of their time high up in the trees, where their thin, shrill 'zee-zee' calls disclose their presence. Their song is also very 'thin' and squeaky, a repeated 'seeter-seeter' ending in a flourish; it is so high-pitched that some people cannot hear it.

When & where Scottish breeding goldcrests are resident, but large numbers of immigrant birds sometimes arrive on the east coast in autumn.

SPOTTED FLYCATCHER

Muscicapa striata
Breacan Glas

Identification

These flycatchers characteristically sit very upright, on a prominent perch, from which they make short flights after insects. They have flattish, broad-based bills, large dark eyes and rather short legs. The sexes are alike; young birds have a scaly appearance to head, back and breast. Its flight is fluttering and erratic, with rapid twists and turns.

Habitat This species frequents woodland edges and glades, parks, large gardens and farms. It is most often found in deciduous or mixed woods. It nests in holes, on ledges or among ivy and will use open-fronted nest boxes or half coconut shells.

Voice The spotted flycatcher's voice is thin and squeaky: its call is a scratchy 'tzee' and its insignificant song comprises only five or six notes.

When & where Spotted flycatchers are widely distributed on the mainland and inner islands. Dependent on flying insects, they are relatively late arrivals in spring.

PIED FLYCATCHER

Ficedula hypoleuca
Breacan Sgiobalt

Identification The male pied flycatcher is the only small woodland bird with a black back and white underparts; it also has a broad white wing patch and a smallish white area on the forehead. Females are olive-brown above and less pure white below and they have a smaller wing patch. Both sexes have white outside edges to the tail; these and the wing patches show well in flight. After alighting they often flick their wings and move their tails up and down.

Habitat This species frequents deciduous woodland, usually near water and where oaks are present.

Breeding Numbers vary, as does breeding success; in wet years the caterpillars essential for rearing young are often washed off the trees, becoming unavailable to the birds.

Voice The pied flycatcher's song is a repeated high two-note 'zee-it', concluding with a trilling warble.

When & where It is most abundant in the Trossachs–Loch Lomond area, Argyll and Dumfries–Kirkcudbright, but occasionally nests further east and north.

LONG-TAILED TIT

Aegithalos caudatus
Cailleach Bheag an Earbaill

Identification A tiny body and disproportionately long, narrow tail are striking features of this delightful little bird. The adults' blackish, pinkish and whitish plumage is unlike that of any other species; young birds lack the pink and have more extensive black on the head. Long-tailed tits are active and acrobatic, sociable and noisy. They move around in family parties or flocks, keeping up a continual, and easily recognisable, conversational chorus of trilling 'sirrup' and low 'tup' calls. Their long tails are very obvious, both in flight and when moving about in trees, allowing them to be readily identified by silhouette alone.

Habitat Long-tailed tits, which build domed nests, are resident in a wide variety of woodland types distributed throughout the mainland, but seem to prefer fairly open birch or mixed deciduous woods.

Feeding They are not regular bird table visitors though they occasionally come to peanut feeders, where they are much less aggressive than other tits.

MARSH TIT

Parus palustris
Ceann-Dubhag

Identification This tit can be distinguished from a coal tit (p.149) by the lack of a white patch on the nape; from a willow tit by its voice, its tiny neat chin patch, and the fact that its black crown is glossy not sooty; and from a male blackcap (p.139) by its bib, much shorter bill and plumper body. Its calls are a loud 'pitchew' and a nasal 'chika-dee-dee-dee'. Its song is a repeated 'shippishippi'.

When & where Local in southeast Scotland.

WILLOW TIT

Parus montanus
Ceann-Dubhag an t-Seilich

Identification It has a larger and less cleanly edged bib than the marsh tit, is much plumper than a male blackcap and lacks the coal tit's white nape patch. Its calls are a wheezy 'eez-eez-eez' or a very high 'zee-zee'.

When & where Locally common only in the southwest, where it occurs in scrubby woodland with birch and alder, often along a river or loch shore or on waterlogged ground.

CRESTED TIT

Parus cristatus
Cailleach Bheag a'Chirein

Identification This is the only small bird with a distinct crest; its speckled crown feathers are elongated to a point and can be raised or lowered, making the crest more, or less, obvious. It also has a distinctive call, a low purring trill, rather like that of a long-tailed tit but lower; this serves to keep members of a family party or small group in contact as they move around, busily searching for insects. They are less gregarious than other tits but occasionally join up with coal tits (p.149) and treecreepers (p.152).

Habitat Crested tits are birds of the pine woods, in summer feeding mainly in the foliage, but in winter often coming down into the shrub layer of heather and juniper. They normally excavate their nesting holes in dead stumps, but use nest boxes in plantations.

When & where Crested tits breed only in the highlands, most regularly in the native pine woods of the Spey Valley, but also now in the Scots pine plantations of the Culbin Forest.

COAL TIT

Parus ater
Cailleachag a' Chinn-Duibh

Identification The coal tit, smallest and dullest in colour of the three common tits, has a bold white patch at the nape which distinguishes it from all other black-capped tits.

Feeding Coal tits are low in the pecking order, giving way to both great and blue tits; they tend to make fleeting visits to bird tables to grab a peanut and take it away to eat in peace, rather than to feed on the spot and risk a confrontation. In their natural woodland habitat they sometimes store seeds behind tree bark.

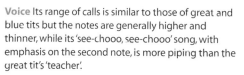

Voice Its range of calls is similar to those of great and blue tits but the notes are generally higher and thinner, while its 'see-chooo, see-chooo' song, with emphasis on the second note, is more piping than the great tit's 'teacher'.

When & where The coal tit is one of the few species which prefer conifer woodlands and has spread as a result of recent afforestation; it is now absent only from the Northern Isles and Outer Hebrides.

BLUE TIT

Parus caeruleus
Cuilleachag a'Chinn Ghuir

Identification One of the most frequent and easily recognised garden visitors. Young blue tits, which are yellowish-green, might be mistaken for warblers at first glance but have much shorter bills and a darkish line through the eye and round to the nape.

Feeding They feed mainly on trees, from ground level up to the top, and only occasionally on the ground. In their natural woodland habitat the hatching of the young is synchronised with the main emergence of moth caterpillars, an ideal food for young chicks. In winter, blue tits often flock with other tits in woodland.

Voice The blue tit's song is a rapid, liquid trill, starting off with a few single 'tsee' notes. It has a variety of calls, though not as many as the great tit, the commonest being a slightly wheezy 'tsee-tsee-tsee-tsit'.

When & where Blue tits are often very tame and will come to nut feeders fastened to a window pane. They inhabit woodland of all kinds and are absent only from islands where such vegetation is scarce.

GREAT TIT

Parus major
Currac-Baintighearna

Identification This is the largest of the tits and one of the most colourful of garden visitors. Young birds have a brownish crown and a paler, rather washed-out appearance.

Feeding They often feed on the ground, foraging among dead leaves for spiders, worms and insect larvae, and are adept at using their feet when feeding, not only to hold down large seeds, but also to pull up strings holding peanuts.

Voice The great tit has a very varied repertoire of calls, most of which have a somewhat metallic sound, including a ringing 'tinktink', and 'tui-tui' repeated several times and rising towards the end. Its most easily recognised song is a loud 'teacher-teacher'. Song starts in January and continues until about June.

When & where Great tits frequent all types of woodland, scrub and hedges, but show a preference for deciduous trees; they are scarce on many of the islands.

TREECREEPER

Certhia familiaris
Snàigear

Identification This is the only small bird which feeds by climbing up tree trunks and branches; it usually spirals up a tree, searching cracks in the bark for insects, then flies down to the base of the next one. Its rather long down-curved bill and the stiff pointed feathers in the centre of its graduated tail also distinguish it from all other small brown birds.

Habitat Treecreepers frequent all types of woodland, though they are more abundant in deciduous and mixed woods than in conifer plantations. In parks and estate woodlands with soft-barked redwood trees, it is easy to spot a treecreeper's roosting place: an oval hollow in the bark, with a white splash of droppings at its lower end. They often nest behind a loose slab of bark, or in the crack of a broken branch.

Voice It is unobtrusive and often first draws attention to itself by its high thin 'tsee' and 'tsit' calls. Its song, a series of 'tsee' notes ending in a faster flourish, is so high-pitched that many people have difficulty in hearing it.

When & where Widely distributed virtually everywhere there are trees.

JAY

Garrulus glandarius
Sgraicheag

Identification Jays glimpsed in flight at the edge of a wood are easily identified by their conspicuous white rump, contrasting with black tail, pinkish back and rounded dark wings with a vivid blue shoulder patch. Jackdaw-sized, they look predominantly pinkish brown at rest, with white and blue patches on their dark wings. Their black and white crown feathers are often raised to form a crest.

Feeding Jays have a varied diet, including insects, acorns (which they sometimes store), eggs and young birds. They feed both on the ground and in trees, hopping about with jerking tails.

Voice The usual call is a loud and very harsh 'kraak', given frequently as small parties move about in the woods.

When & where Although found mainly in deciduous woods they may also be seen along the fringes of conifer plantations. Jays are most abundant in the southwest, Borders and Argyll and in the northern half of the central lowlands.

MAGPIE

Pica pica
Pioghaid

Identification The magpie's very distinctive pied plumage and long wedge-shaped tail make it impossible to confuse with any other bird. Its flight is direct and rather slow and the birds often fly in single file. A rapid, harsh chattering 'cak-cak-cak' is the commonest call; the species has no recognisable song.

Feeding This species is still regarded as a pest as it eats the eggs and young of other birds. Grain, seeds, insects and worms are also taken. The birds walk and hop around, usually with the tail a bit raised.

Breeding Rather surprisingly, in view of their long tails, magpies build domed nests, usually siting these untidy-looking structures high up in a tree or hedge.

When & where Magpies are most abundant in the Clyde–Forth valley and along the Aberdeen–Kincardine coastal strip and are most likely to be seen in parks, around suburban gardens, and on farmland with unkempt hedges and rough grassland. They are very local on low ground elsewhere in Scotland and are absent from the highlands and islands.

CHOUGH

Pyrrhocorax pyrrhocorax
Cathag Dhearg-Chasach

Identification Its red legs and bill distinguish this jackdaw-sized bird. Choughs are sociable and often indulge in aerial acrobatics.

Habitat This species nests in caves and derelict buildings and feeds on invertebrates in the soil, especially on cattle-grazed grassland.

Voice They sound rather like jackdaws (p.156) but their 'keeow' call is higher, more drawn-out and more musical than the jackdaw's 'kya'

When & where Most of the small population of choughs is on Islay, around the Rhinns and the Oa, with a few on Jura, Colonsay and the Kintyre peninsula.

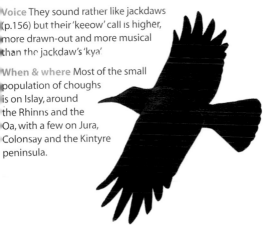

JACKDAW

Corvus monedula
Cathag

Identification Jackdaws, the smallest of the 'black crows', are readily identified by the pale grey area around the sides and back of the head. Their underparts are dark grey, rather than black, and young birds have a brownish tinge. Adults have distinctive pale grey eyes. The jackdaw is a neat and alert bird, which walks about jauntily, often glides near cliff faces, and performs aerial acrobatics.

Feeding Its varied diet includes insects, cereals, peanuts and scraps at bird tables. Jackdaws are very gregarious and form large flocks.

Breeding They often nest socially, with many pairs occupying adjacent holes.

Voice A fairly sharp, high-pitched 'chak' is the usual call, much used in flight to keep the members of a flock in contact and run together when excited to give a chattering 'chaka-chaka-chak'.

When & where They are widespread, occurring in towns, wooded farmland and on hillsides and coastal cliffs.

ROOK

Corvus frugilegus
Ròcais

Identification Rooks are glossy black all over, with loose feathering on the 'thighs' which gives them a 'baggy breeks' appearance. Adults have dagger-shaped greyish bills, at the base of which is a bare whitish patch; young birds have black bills and fully feathered faces, making them look more like carrion crows (p.158). Rooks are very gregarious and are nearly always in flocks. Winter roosts may hold thousands of birds.

Feeding Rooks feed over open ground, often with jackdaws, and eat cereals, worms and insects.

Breeding A small wood or strip of trees seems usually to be preferred for breeding; the untidy stick nests are sited high in the trees.

Voice Less harsh and croaking than a carrion crow's; its calls include a varied range of 'kaa' and 'kaw' notes.

When & where This species is very closely associated with farmland and is consequently absent from most of the highlands and islands.

CARRION/HOODED CROW

Corvus coron...
Feanna...

Identification Carrion crows are glossy black all over and can be distinguished from rooks (p.157) by their heavier black bills with no bare patch and the absence of shaggy feathering around the upper legs. Hooded crows are pale grey on the back and belly. Hybrids vary from dark grey on back and belly to almost black. The crow's call is a deep hoarse 'kraak', usually repeated several times. Voice, and the square-ended tail, are useful in distinguishing carrion crows from ravens (p.159) in the highlands and islands.

Habitat Crows occur in a wide variety of habitats. They nest solitarily on trees and cliffs, and take a wide variety of foods. They regularly breed in towns and scavenge around rubbish tips.

When & where Carrion and hooded crows are more or less separated geographically: east of a line from Speymouth to the Inner Solway carrion predominate; west of a line from Bettyhill to Bute most are hooded; and in the north–south band in between both are present and interbreeding occurs.

RAVEN

Corvus corax
Fitheach

Identification
Their flight is direct and powerful and they frequently perform aerial acrobatics, rolling, gliding and diving steeply with folded wings; the wedge-shaped tail tip shows well in flight.

Habitat Ravens frequent hill country, moorland or sea cliffs, feeding mainly on carrion. Where the supply of carrion decreases in a traditional raven breeding area, the birds often cease to breed and may desert the area altogether. Although generally found in thinly inhabited parts of the country, they come close to man on some of the islands to scavenge on garbage tips.

Voice Ravens call often in flight and their voice is distinctive, a throaty 'pruk-pruk', deeper and less harsh than the 'kraak' of the much smaller carrion crow.

When & where They breed early and remain in family parties through the summer, sometimes joining up to form larger flocks in winter. They are most common in the west and northwest highlands and the islands, more thinly scattered in other upland areas.

STARLING

Sturnus vulgaris
Druid

Identification In winter, buffish spots give the quarrelsome starling a speckled appearance, but these literally wear off as the feathers become worn and by summer have almost vanished. Newly fledged young birds are unspotted mouse-brown with a whitish chin patch.

Feeding Starlings are opportunists, nesting in the roofs and walls of buildings, roosting in vast numbers along wires in city streets and under bridges. They feed readily at bird tables, on rubbish tips and in the open, bustling about with half-opened bills as they probe the ground for insect larvae. They are very social birds, both feeding and roosting in large gatherings.

Voice The starling's song is a lively rambling warble, usually including whistles, gurgles and clicks, and it is a good mimic – curlew calls and goose noises often indicate the starlings' arrival. Its 'conversation' includes a grating 'tcheerr' and a harsh alarm scream.

When & where Widespread and common.

WAXWING

Bombycilla garrulus
Gocan Cireanach

Identification The waxwing's crest is striking, even if the bird is seen only in silhouette, when its dumpy shape and short tail are also apparent. In good light the closed wing shows as boldly marked with white, red and yellow towards the tip, and the yellow tail tip is obvious. Waxwings in this country tend to be rather silent, but occasionally give a quiet trill, usually in flight.

When & where This is one of the so-called irruptive species which visit this country, irregularly and in varying numbers, because the crop of berries, especially rowans, in their Scandinavian breeding area has proved inadequate to sustain them over the winter. Waxwings are gregarious and usually move around in flocks, the size of which depends upon the scale of the irruption. In some winters several hundred appear, in others none at all. They are most likely to be seen on berry-bearing shrubs, such as cotoneaster, and often stay around the same area for several days, spending many hours just sitting about.

HOUSE SPARROW

Passer domesticus
Gealbhonn

Identification The bold house sparrow is familiar to everyone who gives birds even a passing glance. The only species with which the male can be confused is the closely related tree sparrow (p.163), which has a chestnut, not a grey, crown. Females and young birds are vaguely similar in appearance to young chaffinches (p.164), but are much browner and have shorter tails, without any white and only the faintest suggestion of a wing bar. They are busy, bustling birds which indulge in frequent squabbles. This species is gregarious at all times and often joins up with finch flocks in winter.

Feeding House sparrows feed mainly on the ground – in the countryside on grain, in towns on scraps, and in gardens on seeds and insects. They can be a nuisance in gardens as they have an annoying habit of eating flowers.

When & where This is the species most closely associated with man; it occurs throughout the country, but only in the vicinity of buildings and cultivated fields.

TREE SPARROW

Passer montanus
Gealbhonn nan Craobh

Identification Tree sparrows differ from house sparrows (p.162) in having a chestnut crown and a black spot on the cheek and in the sexes being alike. They are also smaller and slimmer, with smaller and tidier bibs. Young birds have smudgy marks in place of the black spot and bib. It feeds on grain and seeds and is sociable, often occurring in small flocks and occasionally joining up with finches and house sparrows in winter.

Breeding Generally a hole-nester, it readily occupies nest boxes and sometimes breeds semi-colonially, with several pairs building close together in 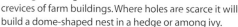 crevices of farm buildings. Where holes are scarce it will build a dome-shaped nest in a hedge or among ivy.

Voice The tree sparrow's flight call, a rather hard and hoarse 'tek-tek', is distinctive, as is one of their other calls, a short metallic 'chik'.

When & where A bird of low cultivated ground with hedges or scattered trees, tree sparrows are most abundant in the Lothians and Fife.

CHAFFINCH

Fringilla coelebs
Bricein Beithe

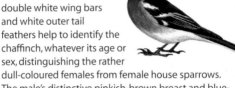

Identification Its double white wing bars and white outer tail feathers help to identify the chaffinch, whatever its age or sex, distinguishing the rather dull-coloured females from female house sparrows. The male's distinctive pinkish-brown breast and blue-grey head are brighter in spring than in winter.

Habitat The chaffinch has adapted to many different habitats, from town gardens to hillside birch woods – wherever there are trees and shrubs.

Feeding It feeds mainly on seeds in winter, but on insects in summer. It also readily comes to bird tables and picnic places.

Voice A loud 'pink' or 'chwink' and in flight a rather subdued 'tseeip' are the chaffinch's most common calls. Its song is lively and relatively short, accelerating as it descends the scale and ending with a fast up and down flourish – 'cherwit-teeoo'.

When & where Scottish chaffinches remain close to their nesting area all year, but in winter immigrant birds from Scandinavia arrive in search of food.

BRAMBLING

Fringilla montifringilla
Bricein Caorainn

Identification Bramblings look like pale chaffinches (p.164) at first glance, but can be distinguished by their white rumps and much paler bellies. The male's head and back are brownish in winter, becoming black by spring. Females have dark markings on their grey-brown heads.

Feeding Bramblings often flock with other finches, feeding over farmland and in woodland, especially on beechmast. In severe weather they sometimes come into gardens for seed but seldom land on bird tables.

Voice Bramblings have a distinctive hoarse, metallic 'tsweek' call; in flight they give a repeated 'chuk-chuk-chuk'.

When & where Variable numbers arrive from Scandinavia to winter in this country; in some years flocks several hundred strong are quite common. They are most likely to be seen on the eastern side of the country.

GREENFINCH

Carduelis chloris
Glaisean Daraich

Identification Bright yellow patches on the wings and sides of the tail distinguish the greenfinch in all plumages. Females are brownish-green, faintly streaked, and young birds brown, heavily streaked.

Habitat Greenfinches were until quite recently birds of farmland, nesting in hedges and wintering on the grain and weed seeds around stackyards. Since stacks vanished from the farming scene this species has become much more a garden bird.

Feeding It is partial to peanuts and, although primarily a ground feeder, has learnt to hang on nut baskets. It also likes sunflower seeds.

Voice Calls include a long drawn-out nasal 'dzeee' and, in flight, a trilling musical 'chi-chi-chi'. Greenfinch song is a rather monotonous warbling twitter, including wheezy 'dzeee' notes, delivered from the top of a tree, or in a slow-flapping song flight.

When & where Greenfinches are widespread in all lowland areas of Scotland.

GOLDFINCH

Carduelis carduelis
Deargan Fraoich

Identification Adult goldfinches of both sexes are unmistakable with their bright red faces and black and white heads. The flight pattern of conspicuous yellow wing bars and a white rump distinguishes this species at all ages. Young birds are streaky grey-brown, with no distinctive head pattern, and when at rest can be confused with other species. Goldfinches have a flitting, dancing flight and a liquid, trickling 'tswitt-witt-witt' call, which is incorporated in the canary-like twittering song. They are not particularly shy and readily perch in the open.

Habitat This species is found wherever there are patches of weedy ground, on neglected farmland, along roadsides and in gardens.

Feeding Goldfinches are remarkably agile little birds, able to hang at thistle and ragwort heads and to pull in a stem with their bills and then use their feet to anchor it as they feed.

When & where They are found, but are not numerous, from the central lowlands southwards.

SISKIN

Carduelis spinus
Gealag Bhuidhe

Identification Smaller, neater and with more streaks than the greenfinch (p.166), this species too has yellow wing bars and sides to the tail. Male siskins have a distinctive black cap and chin and all ages are much more yellow/white below than greenfinches. They are lively and restless, flying in bounding loops and often hanging upside down when feeding.

Voice Their usual call is a wheezy 'tsooeet' and their varied twittering song, often ending in a long wheezy note, is given both in display flight and when perched.

When & where Siskins frequent coniferous woodlands during summer, where they feed on cone seeds, often high up. In winter they move to birch woods and riverside alders, where they may flock with redpolls and also come into gardens for peanuts. This is one of the species which has spread as a result of afforestation. It is now widespread over most of the mainland and several of the inner islands, but breeding pairs are scarce in parts of the central lowlands and near the east coast. Winter flocks, some of which are of immigrants, visit areas in which siskins do not breed.

TWITE

Carduelis cannabina
Gealbhonn-Lin

Identification

Confusion with the rather
similar linnet is most likely
to occur in winter, when
flocks of twites forage on
farmland in low-ground areas where they do not
breed; twites show less white on the sides of the tail.

When & where Breeds on hillsides and coastal moorlands, mainly in the islands and northwest highlands.

LINNET

Carduelis flavirostris
Riabhag Mhonaidh

Identification Male linnets
lose much of the red on
crown and breast in winter.
The whitish patch on the
wing helps in distinguishing
females and young from the
larger yellowhammer (p.175)
and reed bunting (p.176), which have white outer tail
feathers, too. Linnets have a slightly nasal twittering
song and a 'tsooeet' call.

When & where Most abundant south of the highlands,
on rough ground near farmland, occasionally in hedges
and young conifer plantations.

REDPOLL

Carduelis cabaret / flammea
Deargan Seilich

Identification At close range adults are readily identified by their small black chin patch and crimson forehead. Females lack the male's pinkish flush on the breast, and young birds have neither black chin nor red forehead. In flight, redpolls show two rather faint buff wing bars.
Outside the breeding season they are gregarious and move around in flocks from one patch of woodland to another. They are agile and acrobatic and readily hang upside down. This species is the most widespread of the small brown finches. They tend to fly quite high – but their calls enable them to be identified.

Voice It is often their voice that draws attention to redpolls: an almost trilling or buzzing twitter variously represented as 'tyu-tyu-tyu' or 'zz-chee-chee-chee' and easily recognisable with practice.

When & where In summer, redpolls frequent open birch woods, willow scrub, or young conifer plantations, and quite often breed in suburban situations. Many redpolls move south in winter, but some remain.

CROSSBILL

Loxia scotica/curvirostra
Cam-Ghob

Identification Crossbills often first attract attention with their rapidly repeated loud and emphatic 'chip' calls. In flight crossbills look heavy-headed and short-tailed; they usually fly above the treetops. They move around in family parties or flocks, which usually include birds in various plumages: very red adult males, part-red immature males, olive-green and yellowish females, and heavily streaked brownish-green juveniles.

Feeding They feed on cones, usually high in a tree, sometimes hanging parrot-fashion as they wrench off a cone, anchor it with a foot and use their bills to prise out the seeds.

When & where Native Scottish crossbills breed in pine and spruce woods from Tayside northwards. In some years continental crossbills arrive in large numbers from Scandinavia. With many planted forests now mature and producing good crops of cones, some of these immigrant birds have colonised conifer woodlands in many parts of the country.

BULLFINCH

Pyrrhula pyrrhula
Corcan Coille

Identification A white rump, very obvious in flight, and a neat black cap are conspicuous features of adult bullfinches, which are rather dumpy, plump-looking birds. Young birds are pale greyish brown, lack the black cap, and have light-coloured bills. Bullfinches do not have a distinctive song, but their soft piping whistle is often the first indication of their presence, as they seldom descend to the ground and prefer to remain in the cover of trees or shrubs. Bullfinches are more solitary than other finches and are seldom seen in groups bigger than family parties.

Feeding The bullfinch's diet is mainly tree buds and seeds. In the south this species does a lot of damage in orchards by stripping fruit buds but in Scotland this is not such a problem, though bullfinches visiting gardens in early spring often take the buds of flowering cherries.

When & where Their natural habitat is scrubby deciduous woodland but they sometimes frequent conifer plantations at the thicket stage and feed on heather seeds nearby. They are scarce over much of the northern highlands.

HAWFINCH
Coccothraustes coccothraustes
Gobach

Identification This is an elusive and wary bird, which is most likely to be seen as it comes into the open to drink or bathe, when its massive head and bill are very obvious. When it flies between trees, it shows broad white wing bands and a white tip to its tail. Its usual call is a brisk 'tik', which often discloses its presence high in the tree tops.

Feeding Hawfinches feed on large seeds, such as cherry and blackthorn, as well as berries such as rose-hips and hawthorn; their skulls are specially strengthened to support the very strong muscles which enable them to crack even cherry stones.

When & where This species is scarce and local in Scotland. The two places from which it is most frequently reported are the Royal Botanic Gardens in Edinburgh and the grounds of Scone Palace near Perth, at both of which there is a wide range of suitable fruit-bearing trees ensuring an adequate food supply.

SNOW BUNTING

Plectrophenax nivalis
Eun an t-Sneachda

Identification Adult snow buntings are white underneath in all seasons and in flight show a lot of white on wings and tail, which is why they are sometimes known as 'snowflakes'. They call frequently, 'trrip' or 'tyu', and have a dancing flight.

When & where A very few pairs breed on the highest mountains, but many more winter here, when flocks rove around, sometimes joining up with other finches on the fields. They also feed among tidewrack on beaches and marram grass on sand dunes and near the skiers' car parks at Cairngorm and Glenshee.

YELLOWHAMMER
Emberiza citrinella
Buidheag Luachrach

Identification Yellowhammers are buntings with stout bills. Adults have yellow underparts and streaked, chestnut-brown backs; in males the head and breast are bright yellow, while in females they are duller and more heavily marked with dark brown. Young birds are streaky yellowish brown. In flight the combination of chestnut rump and conspicuous white outer tail feathers is distinctive at all ages.

Habitat Yellowhammers are birds of farmland with hedges, or other open ground with bushes and small trees, occasionally nesting along the edge of young conifer plantations. They feed mainly on the ground, taking seeds and grain. They often flock with finches and sparrows in winter.

Voice The yellowhammer's song is often rendered as 'a little bit of bread and no cheese', the emphasis usually coming on the 'cheese'; its calls are a brisk 'tink' and a liquid 'twitup', often given in flight.

When & where Yellowhammers are common in most agricultural districts.

REED BUNTING

Emberiza schoeniclus
Gealag Dhubh-Cheannach

Identification The black head and throat, with white moustache, immediately identify the male reed bunting in summer; the same pattern is less dark and distinct in winter. The streaky-brown females and young birds are rather nondescript, but can be distinguished from female yellowhammers (p.175) by the lack of chestnut on the rump and of any yellowish tinge in the plumage.

Feeding They are more solitary than most other buntings and finches but in winter sometimes join up with mixed flocks feeding on farmland.

Voice The song is a monotonous, deliberate 'tseetsee-tsee-tissick', dropping towards the end; it often sings perched on a tall reed.

When & where Reed buntings are widespread wherever there is marshy ground or reedbeds, on the mainland and many of the islands. In some areas they also frequent hedges, especially in winter, and young conifer plantations.

CORN BUNTING

Miliaria calandra
Gealag Bhuachair

Identification The corn bunting's dumpy shape and short thick bill are helpful in distinguishing it from the similarly streaky brown and much more widespread skylark (p.117) and meadow pipit (p.119). Corn buntings often perch on fence or telegraph wires, or on posts along field boundaries, and have a peculiarly metallic jingling song, like the rattling of keys.

When & where This species was at one time much more abundant on arable farmland but is now regular only along the coastal strip from East Lothian to Buchan and on the islands of Tiree and the Uists, though a few still breed in other areas. It is believed that changing agricultural practices, especially the spraying of cereal crops, has made farmland a less suitable habitat for these birds.

PLACES TO VISIT

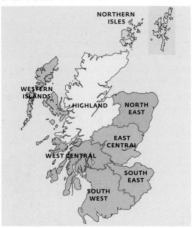

The following sites have been chosen to represent examples of the different habitat types that occur in each of the eight areas shown on the maps; not all habitats occur in all areas. The following abbreviations are used: **LNR** Local Nature Reserve, **NTS** National Trust for Scotland (www.nts.org.uk), **NNR** National Nature Reserve (www.nnr-scotland.org.uk), **RSPB** Royal Society for the Protection of Birds (www.rspb.org.uk), **SWT** Scottish Wildlife Trust (www.swt.org.uk).

South East

The Edinburgh Parks (1) include some excellent inland sites. **Aberlady Bay LNR, Gullane** , and **Gosford Bays** (2) offer a good variety of spring, autumn and winter birdwatching (www.aberlady.org). In summer there are regular boat trips round the **Bass Rock** (3) from North Berwick Harbour, (www.seabird.org). **John Muir Country Park** (4) offers good access to the estuary of the River Tyne (www.eastlothian.gov.uk). Just north of **Coldingham** (5) is **St Abbs Head NNR** (www.nts-seabirds.org.uk), and in Coldstream the **Hirsel estate** (6) supports many breeding birds (www.hirselcountrypark.co.uk). See pink-footed geese flying in to roost at **Gladhouse Reservoir** (7) in Howgate, and walk the **Pentland Hills** (8) to see upland birds. **Thriepmuir Reservoir** and **Bavelaw Marsh** (9) are excellent habitats for water birds.

East Central

The Eden Estuary (1) has good vantage points along its southern shore (www.edenwildfowlers.org.uk). In spring and autumn the Isle of May (2) has large numbers or migrants and breeding seals. Upland birds breed on the heather moorland of the Lomond Hills (3). Loch Leven is one of Scotland's most important wildfowl lochs and the open birch woodland and heather moorland of Vane Farm (4) attract a varied birdlife. Dippers and grey wagtails breed along the wooded stretch of river at Hermitage & Craigvinean (5), 1 mile west of Dunkeld. Follow the network of paths and trails at the Killicrankie RSPB Reserve & Linn of Tummel, Pitlochry (6), and watch breeding ospreys at Loch of the Lowes SWT Reserve, Dunkeld (7). Just 1 mile west of Kirriemuir is the Loch of Kinnordy RSPB Reserve (8). See large numbers of waders in autumn and winter at the Montrose Basin LNR & SWT Reserve (9).

North East

See the large colony of seabirds at **Fowlsheugh RSPB Reserve, Stonehaven** (1) from the cliffs. Extensive woodlands at **Crathes & Drum NTS** properties (2) support many breeding birds. The estate at **Glen Tanar** (3), southwest of Aboyne, has many trails and long-distance paths (www.glentanar.co.uk). The lochs and open birch wood at **Muir of Dinnet** (4) attract considerable numbers of birds in the autumn. Findhorn Bay is a good place to see fishing ospreys and terns, and the pinewoods at **Culbin Forest** (5) support breeding crested tits (www.findhornbay.net). Visit the RSPB reserve at **Loch of Strathbeg, Peterhead** (6), from autumn to spring or watch the passage of seabirds from the lighthouse at **Rattray Head** (7). **Haddo House Country Park** (8) has varied habitats attracting many species. **The Forvie NNR Reserve** (9), just north of Newburgh, is excellent for seabirds.

Highlands

The privately owned **Rothiemurchus Estate**, which lies within the vast **Cairngorm NNR** (1), offers a regular programme of guided walks. Peregrines can be spotted at the **Craigellachie NNR** (2), accessed from the Aviemore Centre, and for an unusually wide range of wildfowl and waders visit the **Insh Marshes RSPB Reserve** at **Kingussie** (3) (www.kincraig.com/rspb). There is a well-known

osprey nesting site at Abernethy Forest (www.nethybridge.com) and in autumn **Loch Garten (4)** has large roosting flocks of waterfowl, including whooper swans. The reserve at **Loch Ruthven, Inverness** (5), is special for its Slavonian grebes. Look out for the crested tit and Scottish crossbill along the paths at the typical highland glen at **Glen Affric** (6), or watch soaring eagles at **Torridon, Kinlochewe** (7), in spectacular mountain scenery. Spot small woodland birds attracted to the exotic gardens at **Inverewe Gardens, Poolewe** (8) (www.swt.org.uk). There are daily sailings in the summer from **Tarbet** to **Handa Island SWT Reserve** (9), West Sutherland, which offers many good viewpoints from which nesting seabirds can be watched. From the viewpoint near the lighthouse at **Dunnet Head** (10), east of Thurso, watch puffins and many other seabirds. The tidal loch and north shore pinewood at **Loch Fleet Reserve, Golspie** (11), are good places for birdwatching throughout the year. Accessed by foot from Kingsteps, the **Culbin Sands RSPB Reserve** at **Nairn** (12) is good for watching hundreds of waders feeding over the flats and saltmarshes, and spotting the visiting birds of prey.

Northern Isles

The remote island of Noss, Shetland (1), can be accessed by small boat from Bressay, or take an excursion from Lerwick Harbour, to visit the huge gannet colony. The accessible storm petrel colony on Mousa (2) can be visited by boat from Leebotten, Sandwick. Watch arctic skuas, kittiwakes and arctic terns bathing in the loch at Loch of Spiggie RSPB Reserve (3) from the viewing points on the north shore. Take a boat or fly in to the Bird Observatory on Fair Isle (4) to watch the 17 seabird species (www.fairislebirdobs.co.uk). Walk from Burrafirth on Unst, to visit the large reserve along the cliffs at the Hermaness NNR (5). The cliffs and moorlands at Fetlar (6) have a good variety of breeding seabirds. Many seaducks, divers and grebes which winter in Scapa Flow can sometimes be watched along the causeways that link the mainland with the islands of Burray and South Ronaldsay, at Churchill Barriers, South Isles, Orkney (7). Take the ferry to the North Hoy RSPB Reserve (8) to see large numbers of great and arctic skuas. Lochs Harray and Stennness (9) are most interesting in autumn and winter, when you can see several thousand wildfowl (www.birdlife.org.uk). From the hide at The Loons RSPB Reserve (10) you can see wetland birds. Marwick Head RSPB Reserve (11) offers good viewpoints on to the breeding ledges packed with

Shetland
Islands

Lerwick

NORTHERN

ISLES

Orkney
Islands

Stromness

Kirkwall

guillemots and kittiwakes. **Point of Buckquoy** (12)
offers a good place for seawatching from July to
October. Watch for breeding hen harriers and short-
eared owls from the hides at **Birsay Moors** &
Cottasgarth RSPB Reserve on the Orkney mainland
(13). Stay at the **North Ronaldsay Bird Observatory**
(14) to watch a variety of breeding birds in spring and
autumn (www.nrbo.f2s.com).

Western Isles

A variety of woodland birds can be found in **Stornoway Woods, Lewis** (1), and breeding great and arctic skuas can be seen on the moorland from **Gress** to **Lolsta** (2). The rocky headland at **Tiumpan Head** (3) is a good place for seawatching. Drive the A865/867 circuit on **North Uist** (4), scanning the moors and hills for hunting birds of prey, and from the causeway at **North Ford** (5) look out over the tidal flats. At **Loch Druidibeg, South Uist** (6), watch for wildfowl and waders. Continue on the A865 to view the large gatherings of mute swans on **Loch Bee** (7). The **Balranald RSPB Reserve** on **North Uist** (8) includes a rocky headland and sandy beaches. Watch the waders feeding at **Vallay Strand, North Uist** (9). The **Clan Donald Centre** (10) on the Armadale estate, Skye, is a good base for observing a variety of birds (www.clandonald.com). The whole of the large and mountainous island of **Rum** is an NNR (11).

West Central

The different habitats in the reserve at **Possil Marsh, Glasgow** (1), make it attractive to a variety of birds. Follow the West Highland Way on foot along the loch shore to **Inversnaid RSPB Reserve, Loch Lomond** (2) (www.loch-lomond.net), or visit the woodlands on the east shore and **Ben Lomond** (3). Follow the narrow **Pass of Leny** at **Callander** (4), through oak woods and alders, past the falls to see woodland birds and moorland species. The vast site of **Queen Elizabeth Forest Park** (5) stretches from the Trossachs to Loch Lomond with a variety of habitats. A maze of tracks run through **Argyll Forest Park** (6). **Islay** (7) offers a variety of birdwatching throughout the year at various reserves, as does **Mull** (8) over the extensive moorlands, young plantations and offshore.

South West

Lying close to the River Clyde, the **Baron's Haugh RSPB Reserve (1)**, includes marshland, meadows, scrub and woodland. See the occasional kingfisher along the stretch of river at the **Falls of Clyde SWT Reserve, New Lanark (2)**. **Caerlaverock NNR** and **Eastpark, Dumfries (3)**, is an important wintering ground for barnacle geese. Extending over a vast area, **Galloway Forest Park** at **New Galloway/Newton Stewart (4)** supports many woodland species, as does the nearby **Wood of Cree RSPB Reserve** at **Newton Stewart (5)**. Seabirds breed on the **Mull of Galloway RSPB Reserve** near **Stranraer (6)**. Much of the **Culzean Country Park**, near **Ayr (7)**, is wooded but wildfowl can be seen on Swan Pond. Two wetland areas at the **Lochwinnoch RSPB Reserve (8)** support wildfowl and small birds breeding in the reedbeds. Follow the nature trails in **Brodick Country Park** on **Arran (9)** to see warblers and moorland birds.

GOING FURTHER WITH BIRDWATCHING

Anyone who becomes really interested in birds will soon want to know much more than can be included in this small volume – about how to identify the rarer species, where and when to visit, and the ecology of each species in different parts of Scotland. No single book provides all this information, but those suggested below cover these aspects. Make contact with other birdwatchers through the SOC and the RSPB (addresses below). A number of bird observatories and ringing groups provide opportunities for more detailed study of birds.

FURTHER READING and ORGANISATIONS

Collins Bird Guide by L. Svensson, P. J. Grant, K. Mullarney and D. Zetterström (HarperCollins, 1999).

Where to Watch Birds in Scotland by M. Madders and J. Welstead (Christopher Helm, 2002).

The Scottish Ornithologists' Club, 25 Ravelston Terrace, Edinburgh EH4 3TP (tel 0131 311 6500). www.the-soc.org.uk

The Royal Society for the Protection of Birds, Waterston House, Aberlady, East Lothian EH32 0PY (tel 01875 871330). www.rspb.org.uk/scotland

INDEX